RESEARCH AND INNOVATION POLICY:
CHANGING FEDERAL GOVERNMENT–UNIVERSITY RELATIONS

EDITED BY G. BRUCE DOERN
AND CHRISTOPHER STONEY

Research and Innovation Policy:

Changing Federal Government–University Relations

UNIVERSITY OF TORONTO PRESS
Toronto Buffalo London

© University of Toronto Press Incorporated 2009
Toronto Buffalo London
www.utppublishing.com
Printed in Canada

ISBN 978-0-8020-9265-6

Printed on acid-free, 100% post-consumer recycled paper with vegetable-based inks.

Library and Archives Canada Cataloguing in Publication

Research and innovation policy: changing federal government–university relations / edited by G. Bruce Doern and Christopher Stoney.

ISBN 978-0-8020-9265-6

1. Higher education and state – Canada. 2. Federal aid to higher education – Canada. 3. Federal aid to research – Canada. 4. Research – Government policy – Canada. 5. Academic-industrial collaboration – Canada. 6. Education, Higher – Canada. I. Doern, G. Bruce, 1942– II. Stoney, Christopher

LC176.R47 2009   379.1'2140971   C2009-904194-4

University of Toronto Press acknowledges the financial assistance to its publishing program of the Canada Council for the Arts and the Ontario Arts Council.

 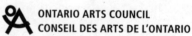

University of Toronto Press acknowledges the financial support for its publishing activities of the Government of Canada through the Book Publishing Industry Development Program (BPIDP).

Contents

Preface vii

Abbreviations ix

PART ONE: CONTEXT, FRAMEWORK, AND DILEMMAS

1 Federal Research and Innovation Policies and Canadian Universities: A Framework for Analysis 3
   G. BRUCE DOERN AND CHRISTOPHER STONEY

2 Pushing Federalism to the Limit: Post-Secondary Education Policy in the Millennium 35
   ALLAN TUPPER

3 Higher Education Funding and Policy Trade-Offs: The AUCC and Federal Research in the Chrétien-Martin Era 59
   CLARA MORGAN

PART TWO: RESEARCH AND INNOVATION POLICY ISSUES, IMPACTS, AND RELATIONS

4 The Granting Councils and the Research Granting Process: Core Values in Federal Government–University Interactions 89
   G. BRUCE DOERN

5 The Canada Foundation for Innovation as
   Patron and Regulator 123
   DÉBORA LOPREITE AND JOAN MURPHY

6 Universities, Commercialization, and the Entrepreneurial Process:
   Barriers to Innovation 148
   PAUL J. MADGETT AND CHRISTOPHER STONEY

7 Intellectual Property, Technology Offices, and Political Capital:
   Canadian Universities in the Innovation Era 172
   MALCOLM G. BIRD

8 Federal Government–University Collaboration in the
   Conduct of Research: Trust, Time, and Outcomes 192
   RUSSELL LAPOINTE

9 The Co-Location of Public Science: Government
   Laboratories on University Campuses 215
   JEFFREY S. KINDER

10 Universities and the Regulation of Research Ethics 242
   KARINE LEVASSEUR

11 Universities and Knowledge Transfer: Powering
   Local Economic and Cluster Development 265
   DAVID A. WOLFE

12 Conclusions: Changing Symbiotic Research
   Relationships: Conflict and Compromise 288
   G. BRUCE DOERN AND CHRISTOPHER STONEY

Appendix: Key Research and Innovation Data 315

Contributors 323

Preface

This book is the product of a collaborative effort by the editors and the contributing authors initiated through the work of the Carleton Research Unit on Innovation, Science, and Environment (CRUISE) in the School of Public Policy and Administration at Carleton University.

Initial draft papers were presented and discussed at the CRUISE Conference on 'Universities and the Powering of Knowledge: Policy, Regulation, and Innovation,' held in Ottawa on 19 and 20 October 2006. Some of the broader conference papers dealing with the larger role of universities were not included in this book so that the book could be better and more explicitly focused on federal research and innovation policies and changing federal government–university relations.

We are especially indebted to the University of Toronto Press's two external assessors, who reviewed the manuscript and made extremely important, constructive, and detailed comments and criticisms. These have resulted in a better, more coherent final product.

In addition to the authors represented in this book, we were fortunate in securing the involvement of leading academics and practitioners, including: Prof Donald Fisher and Prof Kjell Rubenson (University of British Columbia); Prof Saul Schwartz (Carleton University); Dr Sandy Baum (College Board); Prof Robert Johnson (University of Ottawa); Prof Andrea Rounce (University of Regina); Robert Best (Vice President of the AUCC); Prof Ross Finnie (Queen's University); Prof David Cameron (Dalhousie University); Prof Charles Beach (Queen's University); James Turk (Canadian Association of University Teachers); Dr Robert Sauder (Human Resources and Social Development Canada); Prof Jerome Doutriauz (University of Ottawa); Dr Eliot Phillipson (President of the Canada Foundation for Innovation); Paul Dufour

(Office of the National Science Advisor); Dr Peter J. Nicholson (President, Council of Canadian Academies); Prof Susan Phillips and Prof Glen Toner (School of Public Policy and Administration, Carleton University). There was also insightful input from many others who attended the conference or who commented on draft chapters after the conference. Special thanks are owed to Catherine Stoney for her review comments, to Margaret Burgess for her editing of the final manuscript, and to Kimmie Huang for her exemplary support at all stages of this work at both the conference and the book stage. We would like to express our gratitude to all these individuals, who gave unstintingly of their time and expertise. All of them helped to make the book a better product.

Special thanks are due as well for the research funding for this book and related research, which has come from a variety of sources, including the Social Sciences and Humanities Research Council of Canada, the Carleton Research Unit on Innovation, Science, and Environment (CRUISE) at the School of Public Policy and Administration, the Politics Department at the University of Exeter, Industry Canada, the Policy Research Initiative, and Human Resources and Skills Development Canada.

G. Bruce Doern and Christopher Stoney
August 2009

Abbreviations

AAFC	Agriculture and Agri-Food Canada
ACC	Animal Care Committees
ACCC	Association of Canadian Community Colleges
ACOA	Atlantic Canada Opportunities Agency
ADM	Assistant Deputy Minister
AIDS	Acquired Immune Deficiency Syndrome
AG	Auditor General
AUCC	Association of Universities and Colleges of Canada
BNR	Bell Northern Research
BRI	Biotechnology Research Institute
CAER	Centre for Aquaculture and Environmental Research
CANSIM	Canadian Socio-economic Information Management System
CAP	Canada Assistance Plan
CAUBO	Canadian Association of University Business Officers
CAUT	Canadian Association of University Teachers
CCAC	Canadian Council on Animal Care
CEGEP	Collège d'enseignement général et professionnel (College of General and Vocational Education)
CEO	Chief Executive Officer
CFI	Canada Foundation for Innovation
CFS	Canadian Federation of Students
CHST	Canada Health and Social Transfer
CIG	Centre for Integrated Genomics Canada
CIHR	Canada Institutes of Health Research
CIPO	Canadian Intellectual Property Office
CMHC	Canada Mortgage and Housing Corporation
CMAJ	Canadian Medical Journal Association

CMSF	Canada Millennium Scholarship Foundation
COU	Council of Ontario Universities
CRC	Canada Research Chairs
CRCC	Communications Research Centre Canada
CRUISE	Carleton Research Unit on Innovation, Science, and Environment
CSRP	Canada Student Loans Program
CST	Canada Social Transfer
CSTA	Council of Science and Technology Advisors
CUF	Canadian Universities Foundation
CURA	Community–University Research Alliances
DEST	Department of Education, Science and Training (Australia)
DFO	Department of Fisheries and Oceans
DLPP	Dominion Laboratory for Plant Pathology
DM	Deputy Minister
EPF	Established Programs Financing
EPO	European Patent Office
FDA	Food and Drug Administration
FLIP	Federal Laboratory Infrastructure Project
FTE	Full-Time Equivalent
GAAP	Generally Accepted Accounting Principles
GCPSER	Government Caucus on Post-Secondary Education and Research
GDP	Gross Domestic Product
GOCO	Government Owned, Contractor Operated
HIV	Human Immunodeficiency Virus
HQM	Highly Qualified Manpower
HQP	Highly Qualified Personnel
HRDC	Human Resources and Development Canada
ICH-GCP	International Conference on Harmonization – Good Clinical Practice Guidelines
ICMJE	International Committee of Medical Journal Editors
ICR	Income Contingent Repayment loans
ICT	Information and Communications Technology
ILO	Industry Liaison Offices
IOF	Infrastructure Operating Fund
IP	Intellectual Property
IPR	Intellectual Property Rights
IRAP	Industrial Research Assistance Program
ISRN	Innovation System Research Network

JPO	Japan Patent Office
K–12	Kindergarten through Grade 12
KBE	Knowledge-Based Economy
KBES	Knowledge-Based Economy and Society
KPIs	Key Performance Indicators
MTDM	Mechanics of Time Dependent Materials
MOU	Memorandum of Understanding
MRC	Medical Research Council
NAFTA	North American Free Trade Agreement
NBER	National Bureau of Economic Research
NCCU(C)	National Conference of Canadian Universities (and Colleges)
NCE	Networks of Centres of Excellence
NCEHR	National Council on Ethics in Human Research
HHDRP	National Health and Development Research Program
NHRL	National HIV and Retrovirology Laboratories
NINT	National Institute for Nanotechnology
NDP	New Democratic Party
NPM	New Public Management
NRC	National Research Council
NSERC	Natural Science and Engineering Research Council
NSI	National Systems of Innovation
NWRC	National Wildlife Research Centre
O&M	Operation and Management
OECD	Organization for Economic Co-operation and Development
OEM	Original Equipment Manufacturer
OIT	Ontario Innovation Trust
OISE	Ontario Institute for Studies in Education
PBI	Plant Biotechnology Institute
PM	Prime Minister
PMO	Prime Minister's Office
PPSEC	Private Post-Secondary Education Council
PRE	Interagency Advisory Panel on Research Ethics
PRI	Policy Research Initiative
PSE	Post-Secondary Education
R&D	Research and Development
R&DD	Research and Development and Demonstration
REB	Review Ethics Board
RSA	Related Science Activities
S&T	Science and Technology

SARS	Severe Acute Respiratory Syndrome
SBDA	Science-Based Departments and Agencies
SBIR	Small Business Innovation Research
SDTC	Sustainable Development Technologies Canada
SME	Small and Medium-sized Enterprises
SOA	Special Operating Agency
SRC	Saskatchewan Research Centre
SR&ED	Scientific Research and Experimental Development
SSHRC	Social Sciences and Humanities Research Council
STTR	Small Business Technology Transfer
SUFA	Social Union Framework Agreement
SRP	Strategic Research Plan
TCPS	Tri-Council Policy Statement on Ethical Conduct for Research Involving Humans
TPD	Therapeutic Products Directorate
TTO	Technology Transfer Offices
UBC	University of British Columbia
UILO	University–Industry Liaison Office
USPTO	United States Patent and Trademark Office
UW	University of Waterloo

PART ONE

Context, Framework, and Dilemmas

1 Federal Research and Innovation Policies and Canadian Universities: A Framework for Analysis

G. BRUCE DOERN AND CHRISTOPHER STONEY

This book examines the changing relationships between the federal government and Canada's universities as revealed through changes in federal research and innovation policies. As the analysis will show, the federal role vis-à-vis universities has increased in form and importance in the last decade through a new or changed set of programs, processes, and institutions of engagement. Our focus clearly is on Canada but with some relevant contextual reference to broad developments regarding research and innovation policies and universities elsewhere in the U.S., the U.K., and Australia.

The book draws on and contributes to Canadian and comparative literature on science and technology, research, innovation, and related institutional change. This literature includes analysis of commercialization policies through which universities are increasingly expected to be involved in developing new products and processes for the market and also spin-off companies rather than just basic and applied research (Laidler 2002; Wolfe 2003; Atkinson-Grosjean 2006; de la Mothe 2003; Doern and Levesque 2002; Doern and Kinder 2007). Its main contribution in relation to this and related literature is that it is the first to provide a more focused and considered analysis of the ways in which federal research and innovation policies have changed the relationships between the federal government and universities and have impacted on both.

Universities in Canada are creatures of the provinces and are typically examined through the prism of higher education policy. Four recent books cover a range of higher education concerns (Beach et al. 2005; Iacobucci and Tuohy 2005; Jones et al. 2005; Fisher et al. 2005), and others have focused on particular issues such as teaching and academic freedom (Pocklington and Tupper 2002; Horn 1998). This book certainly draws on some of this literature on Canadian higher education policy but neither the Canadian higher education policy literature or the S&T and innovation literature has provided a sufficiently closely linked contemporary account of how universities have been changed and influenced by federal research and innovation policies and how they have in turn sought to lobby the federal government through a variety of institutional arenas.

These changes include the transformation of federal research granting bodies, the creation of new research infrastructure funding organizations such as the Canada Foundation for Innovation (CFI), pressures and incentives to create intellectual property and to commercialize, and the regulation of research ethics, to name only four policy and institutional dynamics we examine.

Our analysis is aimed at a readership that includes both students and practitioners interested and involved in Canadian S&T, research, innovation, and related commercialization policy and its links to universities. The period covered in the analysis is essentially the last twenty years and thus covers mainly the research and innovation policies of the Chrétien and Martin Liberal governments and the Harper Conservative government, but with some important references to the Mulroney Conservative government in the late 1980s and early 1990s as well.

The structure of the book reflects the analytical task at hand. Part 1 of the book deals with the context for, and the dilemmas inherent in, these changing relations and provides a framework for analysis in the book as a whole. Thus, chapters examine the recent evolution of higher education policy and federal–provincial relations and also the lobbying of the key national body that represents universities at the federal level, the Association of Universities and Colleges of Canada (AUCC). This introductory chapter provides the overall four-part analytical framework (see more below), as well as some of the policy background and comparative data on Canada's S&T spending.

Part 2 focuses on key federal research and innovation and related commercialization policy realms, including chapters on: the three research granting councils; the research infrastructure funder, the Canada Foundation for Innovation (CFI); commercialization policies and pressures; intellectual property and technology transfer policies and pressures; university–government collaboration in the actual conduct of research; the co-location of government laboratories on university campuses; the regulation of research ethics; and local economic innovation and cluster development.

The order of these chapters is partly governed by factors such as how old or new the policies or institutional relationships are, the directness or indirectness of federal influence, and the type of policy or governance instrument inherently involved. Thus, the three granting councils come first because they are arguably the most established and recognized federal link with universities, although some federal laboratories and programs such as the Industrial Research Assistance Program (IRAP) and agricultural and mineral labs have an even longer direct linkage (Doern and Levesque 2002; Doern and Kinder 2007). Chapters on local economic and cluster development come at the end of the book in part because they involve newer links but also weaker or less developed ones vis-à-vis the federal government. The regulation of research ethics has strong federal links but also is more unambiguously about rulemaking per se. In between these two poles of ordering is the other range of policy and institutional action, where federal policy is somewhat newer and involves complex mixtures of funding, exhortation, and institutional experimentation.

The editors' conclusions in chapter 12 bring together our views of the contributing authors' assessments of the impacts of federal research and innovation policies on universities and our own overall views with regard to why and how federal research and innovation policies have changed the relationships between the federal government and universities in Canada and the dilemmas inherent in these changed relationships. We offer overall conclusions regarding: a) significant policy impacts; b) complex 'symbiotic relationships,' initially between the federal government and universities, but also with business: symbiotic relationships involve the mutually advantageous association of two or more different institutions increasingly attached to each other or functioning partly within the other; the treatment of business relations and interests extends the conclusions somewhat beyond our federal–university focus per se but is necessary for our overall conclusions in part because federal S&T and innovation policies, as we will see, have sought to establish closer partnered and levered relations between universities and business; c) the dynamic nature of these relationships amidst changing contexts, institutions, and policies; and d) the mutual adjustments needed regarding conflicts and compromises in values and interests that emerge from these impacts and symbiotic relationships.

While the book's purpose is clearly to focus on *federal* research and innovation policies, some aspects of federal–provincial interactions also arise and will be noted. This caveat is needed because obviously some particular federal research and innovation policies may well be based on a reaction to particular developments initiated by particular provinces or may be shaped by the need to ensure that some aspects of federal research support do not offend provincial constitutional-jurisdictional concerns.

Similarly, federal research and innovation policy and the above-noted more explicit commercialization policies often reflect the specific pressures and views of the business community overall or of particular business sectors. Such policies are also tied to business and other political views which favour liberalized market-oriented approaches as a way to structure and encourage research and innovation. Thus, both international and Canadian literature link research and innovation policies to very much varied notions of entrepreneurial or enterprise universities, commerce, and marketization (Geiger 2004; Clark 1998; Marginson and Considine 2000; Bok 2003; Atkinson-Grosjean 2006; Slaughter and Rhoades 2004).

Figure 1.1 provides a simplified initial picture of the federal–university research and innovation policy and institutional relationships to be explored in this book. In an overall sense, it conveys a broad set of quasi-principal–agent relations with the principal overall being the federal government and the agents being ultimately universities and researchers but with many other mediating agencies in between. The 'quasi' aspect of the principal–agent relations designation is needed because typically in principal–agent relations as used by economists, the principal and agent are cast as the individual heads of organizations (ministers, deputy ministers, heads of granting councils etc. Figure 1.1 is crafted also to suggest that the direction of pressure and interaction works in both directions. Through key federal functions, agency modalities, and instruments, federal research and innovation policies impact on universities whose own internal features regarding research and innovation and how it is performed are by no means simple.

In this introductory chapter, we set out our overall framework for examining federal research and innovation policies and federal–university interactions. This framework builds on Figure 1.1 but necessarily has to go into more detail. The chapter is organized into three sections. The first section deals with key definitions. The second section provides a brief contextual picture of the evolution of federal research

Figure 1.1: Federal Government–University Research and Innovation Policy Relationships

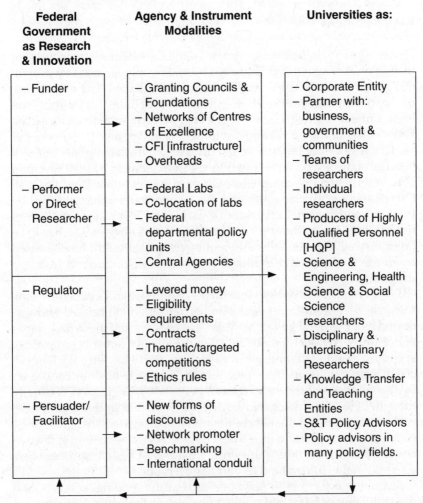

and innovation policy and of basic data on the funding versus the performance of research. The longer third section sets out the analytical framework per se.

Our framework sets out in detail four key analytical elements: the changing overall high-level policy and conceptual discourse; core policy values and ideas; the main policy instruments (taxation, spending, regulation and persuasion and mixes thereof); and institutional and governance change. Brief conclusions then follow.

Key Definitions

Universities and the university 'system' in the Canadian system represent mainly public universities chartered by and mainly financed by provincial governments (Cameron 2003; 2004; 2005a; 2005b). Universities are one part of a higher education system that includes community colleges, but our real focus is on the research-performing university sector. Universities in Canada cannot accurately be called a single system per se. This is not only because of ten diverse provincial systems but also because some see de facto a two-tier system, namely a small group of elite, internationally competitive universities set apart from the rest. Others also see diverse older and newer universities characterized by the local and regional character of individual institutions with their unique histories, traditions, and interests. In this book, notice has to be taken as well of potentially different interests that might be possessed by research faculty and graduate students as distinguished from the university.

Policy refers in this book to statements of purpose and intent regarding research, S&T, innovation, and commercialization enunciated and discussed by the federal government in various ways and in myriad arenas of debate. Such policies mobilize all of the key instruments of taxation, spending, regulation, and persuasion (Pal 2006; Howlett and Ramesh 2003). *Regulation* refers more precisely to 'rules of behaviour backed up by the sanctions of the state' (Doern et al. 1998). Regulation is usually thought of in terms delegated law or 'the regs,' but in real terms, rule-making emerges out of statutes themselves, delegated law, guidelines, codes, and even performance-based targets. As we see below in the discussion of the analytical framework, mixtures of basic instruments are also essential to understand as part of governance.

Research policy refers to policies aimed at the funding, conduct, and dissemination of basic and applied research in the natural, health, and social sciences, but the term is frequently used interchangeably with concepts such as science policy and science and technology policy, and even more broadly as commercialization policy. Basic research typically refers to curiosity-driven research, whereas applied research is more mission-oriented. Research policy, science policy, or overall science and technology (S&T) policy has thus historically been seen as a continuum of activity, with basic research seen as leading to applied research and then to 'development.' The continuum notion is often stretched even further when 'commercialization' emerges. Basic research is typically seen as the preserve of universities, but universities also conduct applied research. Development (linked to our discussion below of innovation and commercialization policies) is typically seen in this basic continuum concept as the preserve of the private sector.

S&T policy also promotes and governs the use of scientific and technical knowledge in public policy and regulation ('science in policy'), where governments need to draw on their internal S&T or the S&T capacities of others to carry out their responsibilities under laws, rules, and international agreements, especially in public interest areas such as environment, health, and safety policy, and regulation (Doern and Kinder 2007).

Innovation policies refer to government policies aimed at fostering the use of the best S&T to produce new and competitive 'first-to-market' products and new production processes, and the innovative organizational approaches and management practices that support these activities (Doern and Levesque 2002; Crocker and Usher 2006). In Canada and elsewhere there is an increasing tendency for governments to subsume S&T and research policies (as defined above) within broader innovation policies in their efforts to maximize the contributions of S&T to the nation's competitiveness in the global knowledge-based economy. Innovation, however, is seen as a non-linear interactive process involving universities, business, and government. Applied development or new measurement techniques can lead to more or different kinds of basic research. Innovation policy thus places considerable emphasis on networks of researchers and on interdisciplinary research.

The overall innovation paradigm has also increasingly been cast in relation to the existence of National Systems of Innovation (NSI) and related concepts of local/regional networks of innovation and 'clusters.' National and local S&T and innovation activity and effectiveness are again seen as being the product of complex non-linear *interactions* among a range of institutions in any national or regional/ spatial political economy, including interactions among universities, corporations, governments, capital markets, systems of regulation, and informed consumerism (Nelson 1993; OECD 1999; Wolfe 2003; Wolfe and Lucas 2004).

Commercialization policies refer to even more targeted efforts to encourage and ensure that innovative new products, processes, and patented inventions are financed, produced, marketed, and successfully sold in Canadian and global markets (Canada 2006). As we will see in chapters 6 and 7, commercialization policies ultimately go well beyond research and science and technology per se and take into account actions tied to risk capital, corporate governance and accountability, and the quality of business management and entrepreneurship. Such policies are also linked to broader notions of the enterprise university and academic capitalism (Marginson and Considine 2000; Geiger 2004). Such commercial behaviour on the part of universities is often a response based on resource dependence. The response is to initiatives taken by the various funding agencies. It can be a voluntary response or it can be a matter of fiscal necessity in that the commercial behaviour may be the result of the withdrawal of different kinds of public funding.

Knowledge as a concept may initially be equated by many with 'research' and the dissemination of research to various users. As we will see, however, knowledge may be seen more broadly as 'knowledge transfer,' where knowledge consists of services rendered and transferred through teaching, research, and service of other kinds. Policy is also increasingly focusing, as David Wolfe's chapter shows, on the importance of 'tacit knowledge,' knowledge that goes beyond theoretical and basic causal knowledge to include experiential knowledge about how to do things and how to think about things in ways that are imparted through both classroom teaching and interpersonal mentoring and coaching (Pavitt 1991; Polyani 1967).

The Evolution of Federal Policy and R&D Funding in the Liberal Chrétien-Martin and Conservative Harper Periods

In addition to the above definitional terrain, our framework for analysis needs to be located within a brief initial look at the evolution of federal research and innovation policy and R&D funding in the Liberal Chrétien-Martin government and Conservative Harper government periods.

In the last twenty years, federal S&T policy has undergone a period of criticism and of both benign and designed neglect, and then a burst of renewed funding that mainly benefited universities (Doern 2006; 2007; de la Mothe 2003). This occurred as new ideas and frameworks for assessing the value of government S&T and innovation policy took hold and as governments in many countries thought seriously about how their citizens could prosper in the knowledge-based economy. Thus, from an early emphasis on a kind of research capacity version of nation building, the rationale for overall S&T policy has increasingly shifted to the promotion of innovation and later commercialization per se. Indeed, the longer term thrust of federal science policy since the 1960s has been to get more of Canada's research and development (R&D) paid for and performed by the private sector.

This was because when federal science policies were first seriously debated in the mid-1960s, the federal government was the main funder and performer of R&D and industry's role was by far the smaller one as a share of total spending. These figures were the polar opposite of the situation in most of Canada's main competitor OECD countries (Doern 1972; Dominion Bureau of Statistics 1970). In the 1970s, this focus on increasing industrial research continued and a federal 'make or buy' policy made explicit that federal 'buying' or procuring R&D was to be the rule for government departments *except* when the research was needed to support policy and regulatory tasks central to the role of government (Government of Canada 1973; 1975). In the 1980s and 1990s, as we will see further below, the policy focus on industrial research continued in the name of innovation policy.

Under the federal Chrétien Liberal government (with Paul Martin as Minister of Finance), the federal granting councils and federal S&T labs and agencies were hit with budget cuts of up to 40 per cent under the deficit-slaying apparatus of the mid-1990s federal Program Review (Swimmer 1996; Canada 2003, Schillo 2003). Since 1997, the federal government, flush with healthy fiscal surpluses, has poured over $13 billion into innovation and S&T funding, but with the great majority of it going to universities via new institutions such as the Canada Foundation for Innovation (CFI), the Canada Research Chairs program, and other funds. The granting councils saw their funding increased, including eventually funds for overheads, but not necessarily in keeping with the growing costs of research or with the growing number of research students and faculty.

Government S&T labs, which often had significant links to universities nationally and regionally, were largely ignored in this re-investment process, at least initially. By 2003, the federal government's own R&D sector did see some increases in funding but these were focused largely in particular federal bodies such as the National Research Council (NRC) and new bodies such as Genome Canada (Schillo 2003). Health Canada's lab in Winnipeg also received a boost in public support after the SARS crisis. However, most of the core labs in the federal science-based departments and agencies (SBDAs) continued to face very tight budget constraints. Moreover, in the interim, the cost of research and research equipment has increased much more than basic inflation rates, and capital investment in the federal S&T infrastructure has not kept pace (Schillo 2003, 5–6).

Meanwhile, these same labs have been required to meet the demands of numerous new policy, regulatory, and monitoring obligations agreed to by the federal government via new international health, safety, and environmental agreements, and new federal laws and regulations. Some labs increased their contacts with universities in order to meet these demands while others reduced them as they scrambled to meet their primary goals (Doern and Kinder 2007).

The various Figures and Tables in the Appendix to the book show the current broad shape of the larger public versus private aspects of Canada's S&T spending in comparison with other countries. Appendix Figures 1 and 2 on overall flows of funding show that business is now by far the main source of R&D funding, and also the largest performer of R&D. Appendix Figures 3, 4, 6 and Table 1 respectively show, however, that Canada still lags behind other countries in terms of government intramural expenditures on R&D as a percentage of GDP, business enterprise expenditure on R&D as a percentage of GDP, human resources in S&T, and levels of patenting. The federal Liberal government aspired to take Canada from 15th to 5th in these global league tables, but there is a very long way to go (Canada 2002; Kinder 2003; Rosenblatt and Kinder 2006).

On the other hand, Appendix Figure 5 shows that Canada leads the world on higher education expenditure on R&D as a percentage of GDP, a result that is directly attributable to the above-mentioned $13 billion in federal spending in the past decade, which was cast broadly as 'innovation' policy but most of which went to universities to support research and the development of related highly qualified manpower.

Within Canada, the data show that business enterprises are now the largest performer of R&D, with 54 per cent of the gross domestic R&D expenditures. Universities and affiliated institutions (such as hospitals) are also major players, performing about 36 per cent of the national R&D activity. The federal government has seen its share of R&D performance decline further over the last decade and now performs less than 8 per cent of the national total (Statistics Canada 2006).

When the Harper Conservative minority government came to power early in 2006, its initial focus on five priorities did not include any reference to research and innovation. However, its policy positions began to emerge, first indirectly in an initial advisory panel report and later more directly in its Budgets and economic updates, and in its 2007 S&T strategy.

The advisory panel reported in 2006 to the Conservatives and some of its ideas were indicative of a Harper government focus on S&T policy, but with a particular emphasis on commercialization (Canada 2006). The expert panel was chaired by business academic Joseph Rotman but also included key industrial experts such as Mike Lazaridis of Research in Motion, a dynamic Canadian innovation company with strong direct links to the University of Waterloo.

As have earlier studies, the panel's report stressed three related aspects of Canada's past and future challenges regarding competitiveness and commercialization. First, it reiterated Canada's weak productivity record compared to the United States, Canada's main market. Second, it pointed out the continuing weakness of Canadian industry's investment in R&D. Third, it noted and applauded Canada's high levels of S&T investment in universities, especially since 1997.

Unlike some studies on commercialization, the Rotman panel argued that the university role in basic research and overall R&D was essential for commercialization, and by implication that the university role in providing highly qualified personnel who are later employed by companies is crucial as a technology transfer conduit. Accordingly, the panel's recommendations focused on building a culture of research in the private sector since the weakness in the commercialization process was in the industry 'demand' side, not the university 'supply' side. The panel's key recommendations included:

- the creation of a business-led Commercialization Partnership Board to advise the Minister of Industry;
- an increase in the business demand for talent through the development of a new Canada Commercialization Fellowships Program, which would include undergraduate, graduate, and post-doctoral fellowships, as well as career interchange fellowships and chairs of research in practice.
- the creation of a Talent and Research Fund for International Study and also initiatives for international students to stay in Canada;
- the creation of a Commercialization Superfund to Address Key Commercialization Challenges (Canada 2006, 8–18).

The panel also made recommendations regarding ways to encourage angel capital investors, small-business partnerships, and business–

university partnership programs already operated by the three granting councils.

Early Conservative Budgets did include increased funding for the three granting councils and for university research overheads (Doern 2006; 2007). They also included a planned study of the granting councils with a focus on their accountability. Commitments were made as well to study the transfer of the management of some government S&T labs to universities. However, it was the 2007 Harper government S&T strategy that brought out a larger sense of the Conservative's policy direction and discourse.

Casting S&T as a central part of the 'Canadian Advantage,' it saw S&T policy as having to contribute to a threefold 'entrepreneurial advantage,' 'knowledge advantage,' and 'people advantage.' It stressed from the outset, however, that 'the most important role of the Government of Canada is to ensure a competitive marketplace and create an investment climate that encourages the private sector to compete against the world on the basis of their innovative products, services, and technologies.' But it also stressed in the next sentence that 'Canada must maximize the freedom of scientists to investigate and of entrepreneurs to innovate' (Canada 2007, 10).

Echoing the earlier advisory panel, the Harper Conservatives' notion of the entrepreneurial advantage centred first and foremost on the 'need for a strong private sector commitment to S&T' (Canada 2007, 9). Thus, among its initial more specific commitments was to:

- Introduce new business-led research networks under the Networks of Centres of Excellence (NCE) program;
- Establish a new Centres of Excellence in Commercialization and Research Program;
- Develop new approaches to transfer knowledge and technologies from universities, research hospitals, and government laboratories to the private sector;
- Create a new tri-council private sector advisory board for the granting councils to provide advice on the implementation of business-driven initiatives (as noted above) (Canada 2007, 94–5).

Following their October 2008 re-election with yet another minority government, the Harper Conservatives and universities and Canadian researchers together faced a vastly different context for S&T and innovation policy. The U.S.-led global credit and banking crisis and the subsequent deep recession globally and for Canada, accompanied by the need for fiscal deficits, the first since 1997, made for great uncertainty regarding future S&T

and innovation policy and funding. The Harper Government's new Science, Technology and Innovation Council, State of the Nation report, and related media reports drew attention to cuts to the granting councils in the January 2009 Federal Budget and to the contrasting case of the new Obama Administration in the U.S. where expanded research support was on offer (Science, Technology and Innovation Council 2009; Ibbitson 2009).

University endowment and research investment and pension funds suffered immediate serious reductions in value of over 20 per cent in late 2008 (Tamburri 2009). The size and urgency of federal deficits meant that extra spending would have to go mainly to industry bailouts and related assistance – in short, a sudden resurgence of industrial policy spending rather than longer term research and innovation policy and investment needs.

While this basic evolutionary picture and accompanying data are important, they are both still broad brush in nature. The framework which follows and our authors' more detailed analyses show that more in-depth assessments are required to examine how federal–university relationships have changed, what funding versus performance actually means when a deeper sense of values and impacts is examined, and when inherent policy dilemmas and conflicts are brought out in a more complete way.

A Framework for Analysis

Table 1.1 presents an initial overall look at our framework for analysing federal research and innovation policies and the changing relations between the federal government and universities. It provides, at a glance, a reasonable conceptual and analytical sense of: 1) high-level policy and conceptual discourse; 2) core policy values and ideas; 3) policy instruments (taxation, spending regulation and persuasion, and mixtures thereof); and d) institutional and governance change.

High-level Policy and Conceptual Discourse

Our policy review above and several chapters in the book show how federal policy has staked out research, S&T, and innovation policy as its main constitutional entry points into university affairs, essentially through the use of the federal spending power but also through a more generalized view that the federal government has the main responsibility for overall economic policy and prosperity in a highly competitive and knowledge-based global economy.

Table 1.1
The Analytical Framework at a Glance

High-level Policy and Conceptual Discourse
- Science policy (policies for science and science 'in' policy)
- S&T policy as linear continuum (basic, applied, development)
- 'Public science' vs. 'government science' and 'academic science'
- Innovation policy as non-linear interactions
- Knowledge-based economy (KBE)
- National Innovation Systems
- Local-Regional Innovation Systems and Clusters
- Changing modes of knowledge production
- The New Public Management (NPM)
- Boundary Organizations

Core Policy Values and Ideas
- the relative independence of academic researchers and the research process from government
- importance of peer review (and merit-based review in infrastructure grants)
- public accountability and value for taxpayer money spent on research
- government and social need for objective and useful social, economic, and policy-relevant research
- the idea of networked and partnership-based research and knowledge-sharing
- commercialization of university research

Policy Instruments and Instrument Mixes
- spending (e.g., research grants and infrastructure grants)
- taxation (e.g., the SR&ED tax credit; tax treatment of charities)
- regulation (e.g., research ethics; accountability and reporting rules)
- exhortation/persuasion (policy studies; benchmarking; general debate and discourse)
- instrument mixes (e.g., levered spending; hard and soft rules; contracts)

Institutional and Governance Change
- governance modes through lobbies such as the AUCC
- arm's-length bodies (changes within and among the three granting bodies)
- the Canadian Institutes of Health Research (CIHR) as restructured networked set of institutes
- newer 'boundary' organizations such as foundations, virtual-networked national centres of excellence, university technology transfer offices
- co-location of federal labs on university campuses
- university spin-off companies
- more overtly innovation-branded universities (e.g., University of Waterloo)
- clusters as local multi-organizational and stakeholder networks

In one sense discourse has shifted in emphasis from science policy to S&T policy to innovation and even to commercialization, but without ever totally abandoning the earlier ascendant concepts. They all tend to be used in the larger policy ether, showing up in speeches, reports, conference discussions, academic seminars, and media coverage.

The concepts of national innovation systems, local-regional innovation systems, and clusters have variously emerged across the last twenty years or more and have been used to capture views about the non-linear nature of research and innovation. Along with concepts such as the knowledge-based economy, these innovation- and commercialization-linked ideas also often imply that one cannot actually have determinative policies about them, but rather only strategies for recognizing and fostering them through networks of interaction. In a sense, therefore, they raise a very old issue about whether research itself can ever be fully 'planned,' either at the level of universities as institutions or of research faculty as employees of such institutions.

Discourse centred around concepts such as the New Public Management tends to be more muted and narrower but it, along with closely related criticisms of bureaucratic hierarchies and the value of networks and partnerships, is important to the debate about research and innovation policy. This is because policy change is seen by many to be crucially dependent on institutional change and new forms of governance. Policy change also creates new defenders of traditional hierarchy and accountability to elected politicians and voters and taxpayers.

The concept of *'boundary'* organizations has also emerged to refer to intermediary agencies inserted between the state and universities and their researchers and decision-makers, or more broadly between 'science and politics' (Guston 2000). This includes longer-standing bodies such as the three main arm's-length granting councils – NSERC, SSHRC, and CIHR – and newer entities such as foundations, the main example of which is the Canada Foundation for Innovation (CFI). Universities, of course, are themselves boundary organizations legitimated by the state but quite independent of it as a part of the para-public sector, several steps removed from the normal state bureaucracy and regular state architecture (Jones 2005).

Other diagnoses of how science and knowledge production have changed have generated their own discourse. For example, Gibbons et al. (1994) had earlier approached science and policy issues from a higher, more abstract level by describing a social transformation of all knowledge-producing institutions. This and related work differentiated between so-called Mode 1 and Mode 2 systems of knowledge production, with Mode 1 roughly referring to familiar academic, hierarchical, and traditional peer review approaches and Mode 2 highlighting emerging networked, flexible, and interdisciplinary approaches with much broader notions of peer review (Gibbons 2000). These authors pointed to the evolution of advanced science systems towards greater heterogeneity, pluralism, and fuzziness of boundaries (between public and private, between knowledge producers and knowledge users, between natural and social sciences, etc.). In particular, the authors describe a shift in the modes of funding and institutional designs of knowledge-producing institutions and a greater emphasis on, and new approaches to ensuring, public accountability for these institutions.

Any understanding of overall policy discourse must also be based on an appreciation that such discourse can often be expressed at a very high level of philosophical expression and also, in the media Internet age, as shorthand sound-bite discourse. Though many aspects of overall discourse, such as innovation, the knowledge-based economy (KBE), commercialization, national and local-regional innovation systems, clusters, new modes of knowledge production, and science policy overall, are underpinned by serious discussion and research, they are often expressed in ways that involve considerable overlap among concepts, and in ways that can be several steps removed from operational policy as expressed in laws, agency mandates, and rules and guidelines. But at the same time these kinds of discourse are important to understand as we look at the more particular research and innovation policies in each chapter in part 2.

Core Policy Values and Ideas

Six core policy values and ideas underpin the research and innovation policy relationship between the federal government and universities. They also produce potential conflicts and barriers and hence limit or configure where partnerships and collaboration may or may not work in specific situations.

The first value that underpins the relationship is *the relative independence of academic researchers and the research process from government*. The implicit contract between the state and the researcher implies that researchers have high degrees of independence. This relates to the choosing of the area or topic, the theoretical perspective, the methodology used, and the conclusions or findings reported. Independence is buttressed by the publication of research subject to peer review processes and an overall view that publicly funded research is in the public domain. However, the independence is always relative rather than absolute in nature. From the point of view of government, the rest of the contract with researchers is that they will carry out this research professionally and competently and also ethically. As we see further below, it also increasingly implies that the research will be policy-relevant and economically and socially useful in defined ways, but very often in broad undefined ways as well.

A second related key value is the *importance of peer review*. Research entails processes involving a community of scholars and researchers that is both Canadian and global in nature. Peer review means review by fellow experts (Canadian and international) in a given discipline or field, with assessments based where possible on the anonymity of the assessors. Peer review is also a part of the selection process in that assessors place a high weighting on the applicant's own publications, which are themselves published in peer-reviewed academic journals and peer-reviewed books and monographs. The analysis of the CFI in chapter 5 also includes the concept of so-called 'merit-based review,' when infrastructure grant decisions are at stake and wider sets of players and expertise participate in review processes.

A third important value is *public accountability and value for taxpayer money spent on research*. Public accountability classically means a public reporting relationship to elected politicians, both in Cabinet and in Parliament, who in turn represent citizens/voters. This involves mechanisms for critical questioning and the need to defend policies publicly and to face and answer questions posed about the three granting bodies and about the spending and arguably about the research overall, or even about individual research grants. These accountability relations can be numerous and increasingly complex.

Accountability in a modern governance era also requires reporting relations which are directed downwards, across, and outwards to Canadian citizens overall – to clients, partners, and network participants – and hence it is quite common to have to speak of 'accountabilities' (Flinders 2002). Accountability *to whom* is also complemented by the question of accountability *for what*? For the ethics of how research is conducted? For how transparent the research is? For how policy-relevant the research is? Value for money is closely linked to the accountability equation, but concerns about efficiency and effectiveness in the use of public funds as well as honesty and probity in the use of tax dollars are also important.

A fourth and more encompassing value is the Federal Government and Canadian society's *need for objective and useful social, economic, and policy-relevant research* in a fast-changing Canadian and global context. There is a sense in which this value gives primacy to publicly funded research as a rightful public good rather than a private good. The state, funded by taxpaying citizens, pays for the research, and therefore society and the economy overall should have access to the research in a useable form. The government itself needs research that is policy and institutionally relevant to ensure that it can carry out its policy, regulatory, and governance tasks in an effective and democratic manner.

With regard to federal policy values and ideas, a picture of both continuity and change emerges as well in Table 1.1. A broad linear concept of overall S&T policy as defined above has certainly been a core idea underpinning federal support for university research (Doern 1972). The three granting councils, as chapter 4 clearly demonstrates, have been the main vehicle for this idea, with research grants in the natural, health, and social sciences being directed to the support of both basic and applied research. Closely tied to this policy value is the importance of developing highly qualified personnel (HQP) or, in the lexicon of the innovation economy, human capital and also social capital.

These core values have not been abandoned but in the last twenty years federal policy has adopted a combined S&T and innovation policy mantra, with the latter informed broadly by the non-linear national and local systems of innovation approach, augmented by general arguments about the ascendancy of the knowledge-based economy. The non-linear notion, as noted in our definitional section, refers to the view that innovation, in the form of new products, processes, and even institutions, is not the product of some linear process of basic research leading to applied research and then to 'innovation' or even 'commercialization,' but rather occurs in multi-directional and interactive networked ways (de la Mothe 2003; Doern and Kinder 2007).

Thus a fifth overall value is growing support for *the idea of networked and partnership-based research and knowledge-sharing*, including the crucial value of the previously mentioned concept of tacit knowledge and new modes of knowledge production (Lundvall 2004; Etzkowitz and Webster 1998; Gibbons et al. 1994). As the chapter by LaPointe shows (chapter 8), such collaboration in the conduct of research centres on complex relationships of trust, time, and outcomes. The chapter by Kinder (chapter 9) also discusses forms of current or potential collaboration, where research conduct and networking is encouraged through the deliberate co-location of government laboratories on university campuses. There have always been important research relationships between federal labs and universities, but recent initiatives have sought to develop these by explicit co-location efforts. These views were also linked to greater scepticism about hierarchical forms of organization in S&T policy overall and in its institutional delivery mechanisms and, indeed, in all modern organizations.

A sixth overall value, though a highly contested one, centres on the *commercialization of university research* (Canada 1999a; Doern 2006; 2007). These ideas, as the chapters by Madgett and Stoney and by Bird (chapters 6 and 7) show, were also complemented by more explicit concerns about the overall competitiveness and productivity of the Canadian economy and the linked research competitiveness of Canada's university system as a national system, including the particular role of Canada's elite globally ranked research universities or those that aspired to such status.

Policy Instruments and Instrument Mixes

The federal government intervenes in and seeks to influence the behaviour of universities and university researchers through the normal set of basic policy instruments and mixtures thereof: taxation, spending, regulation, and persuasion. While these instruments are normally seen as the 'means' of policy action, they are also, in real-world politics, disputed because they imply levels and degrees of coercion which go to the core of how democracy and governance are viewed by different stakeholders (Doern and Phidd 1992; Pal 2006).

On the *taxation* front, corporate tax incentives for R&D are an important indirect element, with impacts on universities and partnered research and innovation funding. At the federal level, there are normal deductions for R&D expenses and also the Scientific Research and Experimental Development (SR&ED) tax credit, which provides funds to smaller Canadian firms that engage in R&D where such firms do not yet have taxable income (Canada 2006; Doern 1995). There has been relatively constant pressure on the federal government to broaden the SR&ED tax credit or to make it less bureaucratic, steps that current federal Conservative government S&T policymakers promise to study and act upon (Canada 2007). Recently, changes to taxation of charities and capital gains have also led to major increases in the charitable donation of stocks to universities for the support of focused research and teaching programs.

With respect to *public spending*, the main instrument used by the federal government, the trends in using the power of the purse to support S&T and innovation policy are more complex than for taxation. The granting councils have been the central feature of the federal spending role for research. However, as the Doern analysis in chapter 4 shows, each of the three granting councils, NSERC, CIHR, and SSHRC, have been subject to greater pressures to develop granting programs that are more targeted and policy-relevant (in effect spending with rules attached). This pressure has come not only from within the federal government but also from entrepreneurial academics within universities. The granting councils, as noted above, were hit with expenditure cuts in the mid-1990s (and even earlier), but since federal budgetary surpluses emerged in 1997 they have done relatively well in the federal budgetary competitive stakes (Lopreite 2006; LaPointe 2006; Murphy 2007).

Up to $11 billion of the overall S&T and innovation funding increase found its way directly or indirectly into the university realm during the Chrétien-Martin Liberal era. As Morgan's analysis in chapter 3 shows, this ascendancy of the federal role in university research funding was very much the product of direct lobbying by the universities through the AUCC, and crucially as well by key university Presidents such as UBC's Martha Piper. However, it also required a receptive senior Cabinet supporter and in this case it was undoubtedly Finance Minister Paul Martin, with the full backing of Prime Minister Jean Chrétien.

An early precursor to the innovation-centred support had come in the late 1980s, when funding began for the Networks of Centres of Excellence (NCE) program (Atkinson-Grosjean 2006). This innovation was in turn partly a federal response to the mid-1980s establishment of centres of excellence programs and networks by Ontario and Quebec. But the real burst of federal funding and program and institutional experimentation came in the successive establishment of the Canada Foundation for Innovation (CFI), the Canada Research Chairs program, the Millennium Student Scholarships, and federal spending support for research overheads. Spending initiatives via Genome Canada and Sustainable Development Technologies Canada (SDTC) were also important (Doern 2006; AUCC 2005).

As chapters 8 and 9 show, an increased federal presence was also emerging through federal laboratories (NRC and other departmental labs), where greater cooperative joint research was being encouraged, and in the co-location of some federal labs on university campuses (Doern and Kinder 2007).

As we have already noted earlier in the chapter, it is by greater use of the *regulatory* instrument that other significant policy interventions have occurred, so much so that one must increasingly speak of the universities as a regulated sector or even, despite the contentiousness of the term, as a regulated 'industry.' Contemporary regulatory analysis stresses the growing range of rulemaking activities carried out by the state on its own or with stakeholders. Thus it should be no surprise that rulemaking has become more extensive and embedded as various elements in society exert their views, politically and economically, as to how they think universities should behave and how access to, and the use of, knowledge should be governed. It has also taken the form of multi-level regulation (Doern and Johnson 2006), as both levels of government, and indirectly local governments as well, use rules, guidelines, and performance standards to shift universities and knowledge in the direction of varied notions of the public interest and of private interests as well.

With respect to *federal regulation*, several chapters in this book make it clear that rulemaking of different kinds and degrees has increased in significant ways. In many cases the federal government is reluctant to call it regulation per se because of jurisdictional niceties, so it is often carried out through arm's-length bodies and boundary institutions (see further discussion below of institutional change).

One major change, as Lopreite and Murphy's account of the CFI in chapter 5 makes clear, is the CFI's requirement that all universities (and research hospitals) bidding for infrastructure funds must submit a strategic research plan. This in turn changes decision-making processes within universities, many of which previously tended not to have such strategic plans. Some universities did of course have research plans or strategies, but the CFI made these much more crucial for getting infrastructure money. The CFI cannot tell the university what their planning processes should include, but it can and does influence how such processes evolve through infrastructure applications requirements. The CFI also has rulemaking elements in its mandate simply because its spending comes 'with rules,' namely requirements to bring money from other funding partners (public or private) in order to get CFI money.

Chapter 3's account of the role of the AUCC shows that it, too, has become a vehicle for rulemaking in the form of broad agreements between the federal government and Canada's universities (and community colleges) on overall financial accountability and responsibility matters tied to the receipt of federal grant funding (Canada 2002).

Similar agreements have been brokered regarding rules about research ethics.

However, chapter 10's assessment by Levasseur of the ethics-rules regimes shows how the earliest crucial forms of ethics rulemaking were a form of self regulation by researchers and thus embedded in professional norms about the conduct of research. The Levasseur analysis shows how extended granting council and other rules in turn force universities to become internal regulators, and not just managers, of their own faculty and research students. It also shows how federal drug regulators regulate research done in universities by requiring pre-clinical testing of drugs at an early stage of research. Concerns also arise within the research community about regulatory compliance and the effects of these rules when the dominant model of ethics rules is centred on medical health research, the rules for which are often inappropriate and excessive for research in the natural and social sciences. On numerous occasions those being regulated have complained about rigid 'one size fits all' approaches to rulemaking and about excessive red tape.

Other areas of embedded rules and regulatory impacts have come from intellectual property (IP) rules. As Bird's analysis in chapter 7 shows, IP rules in the form of patents derive from national law as administered by the Canadian Intellectual Property Office (CIPO). Such laws do not differentiate universities or faculty members from other inventors or patent applicants. But they do indirectly require universities to internally regulate how they will handle IP matters and core ownership questions regarding such property as between the university as a corporate entity and its employees and students.

Concerns have emerged about the overall effects of granting intellectual property (IP) rights over human genetic materials in the health sector (Canadian Biotechnology Advisory Committee 2006; Health Canada 2007). A more recent particular manifestation of these concerns has arisen in relation to control over access to patented genetic diagnostic tests. Several high-profile cases triggered these concerns when patent holders exercised their rights in ways that many regarded as harmful to the provision of health services and to innovation overall in that they impeded access to patented genetic diagnostic tests in areas of public health research conducted at both universities and hospitals. A related concern centred on the potential for patent holders to exact excessive rents by charging high prices for their products or services, thereby imposing heavy cost burdens on the health system not only for patients but also for health researchers using these genetic diagnostic tests.

The use of *exhortation and persuasion* as a policy instrument in federal research and innovation is also growing. For example, the expansion of IP rights has been actively fostered by the federal government (and some provincial governments) as a better indicator of real innovation occurring or potentially occurring than simply doing R&D or using R&D rates as an indicator of progress (Doern and Sharaput 2000). Universities have been increasingly exhorted to join the IP club so that taxpayers get better overall economic value from the university research being funded at the federal level (Canada 1999a).

Federal exhortation is also endemic in the entire broad array of high-level policy and conceptual discourse referred to in the first part of the analytical framework. Such broad kinds of discourse and changing policy labels are all, in their own way, intended to coax and cajole various stakeholders towards a new or updated view of what needs solving and about how generally to think about it. This exhortative discourse can, of course, send mixed and even confusing signals, but there is little doubt that it percolates to and through universities as well as other stakeholders in the research and innovation policy process.

Finally, the federal government has sought to coax and cajole through performance assessment processes and benchmarking. The aspirational pronouncement that Canada needs to move from 15th to 5th in the overall global S&T and innovation league tables is an example here. Establishing a Canadian Council of Learning is hopefully also a way of eventually ensuring that Canada's universities perform well at a globally competitive level, something that can only be judged if performance is openly assessed on a comparative global basis (Canadian Council on Leaning 2006).

Institutional and Governance Change

It is impossible to review the above-noted factors such as policy and conceptual discourse, policy values, and policy instruments and changes without simultaneously noting the role of institutional and governance change. Thus we have already referred to and hinted at key institutional features and examples of change. In this final portion of the framework we draw together some of these earlier observations in a summary form.

As Table 1.1 shows, institutional analysis needs to reference governance changes through the role of national lobbies such as the AUCC. We have stressed the AUCC's role as lobbyist, but also as a broker and intermediary body. One can easily cast the AUCC as just another interest group, but this would be somewhat inaccurate in that, though it exists mainly to influence the federal government, it must do so in a cautious and non-strident way because, in the final analysis, it represents a complex group of institutions that is a part of the provincial para-public sector and among which there are often conflicting views about higher education and research policies. We have also noted the AUCC's governance role regarding agreements between the federal government and universities.

The CFI has also been cited above in several different contexts, but in institutional terms it remains a leading example of the federal use of the foundation model, a boundary organization both praised for its largely non-political decision-making and infrastructure funding approach, but also criticized as an arm's-length step too far – in other words, an organization dispensing taxpayers' money (its initial and only endowment) without being accountable in the 'normal' way to Parliament and elected politicians (Aucoin 2003). Though a Chretien Liberal government creation and institutional innovation, the Harper Conservative government also sees some value in this foundation mode with its plan for the creation of a new Post-Secondary Education Infrastructure Trust. On the other hand, the Conservatives dissolved the Canada Millennium Scholarship Foundation; thus, the foundation as boundary organization may be viewed more critically as compared with its heyday in the late 1990s under the previous Liberal government.

A further key institutional change under the federal column in Table 1.1 is the Canadian Institutes of Health Research (CIHR). As recent analysis has shown, the CIHR became in 2000 the successor to the former mainly traditional discipline-structured Medical Research Council (Murphy 2007). Its main distinguishing feature was its several 'virtual reality' institutes, created to deal in a more multi-disciplinary way with applied health problems (such as aging, and Aboriginal health) (Canada 1999b; 2005). These were assembled as research networks and functioned to complement the funding that would still continue for traditional grant funding realms centred on traditional medical and health research disciplines.

As Doern's analysis in chapter 4 shows, the other granting councils, NSERC and SSHRC, were also being influenced by the previously noted pressures to create more targeted policy-relevant research, pressures which suggested overall that the federal government wanted its granting bodies to be somewhat *less* arm's-length realms of research and instead realms more fully engaged with research useful to governments and to Canadians as a whole (Lopreite 2006; LaPointe 2006).

Other research on universities has also exhibited varied ways of thinking about how this core knowledge institution has been changed by parallel pressures to collaborate with government and industry, both because of inherent interest in particular problems and puzzles and also because of the need to garner funding (Atkinson-Grosjean 2006; Boden et al. 2004). This literature has raised any number of practical and value dilemmas as universities became involved in variously cast kinds of public science and private interest or in a broad 'triple helix' of complex relationships (Etzkowitz and Leydesdorff 1997). This relationship is at the centre of the 'symbiotic relationship' referred to by Wolfe and also explored in chapter 12.

For example, the analysis by Atkinson-Grosjean looks at the federal Networks of Centres of Excellence (NCE) program, the federal precursor for the network model and for levered money in Canada, and links it also to networks of excellence established in Ontario in the mid-1980s. She employs a framework that focuses on the public–private dimension and also on the 'basic' versus 'applied' research dimensions, examining the spaces and interstices across these two pairs of domains. She focuses on and expresses serious concerns about the public–private aspects of the NCE networks.

Table 1.1 also notes the more explicit moves to co-locate some federal labs on university campuses (such as the federal Wildlife Research lab at Carleton University). As the analysis by Kinder in chapter 9 notes, the federal Conservative government is interested in looking more closely at alternative management arrangements for federal labs, including potential transfers to universities. He examines the co-location model, and an example of a co-management arrangement, for federal labs located on university campuses that stop short of full transfer but that still provide many of the same benefits.

Also a part of the nexus of institutional and governance change are clusters themselves as local or regional multi-organizational and stakeholder networks, as Wolfe analyses in chapter 11. This kind of change also includes examples of universities creating spin-off companies, and indeed universities such as the University of Waterloo, which brand themselves as newer and very much innovation-centred, company-creating universities.

The analytical framework discussed above is obviously a classification of key elements of analysis for assessing federal research and innovation policies and changing federal government relationships with universities. Some linearity is implied as well through suggestions that broad high-level policy and conceptual discourse lead to and influence particular policy values. There are also links between policy instruments as policy means and overall institutional governance change. But the actual content of policy, including the debate about innovation and networks, suggests strongly the need to appreciate the non-linear and multi-directional nature of the relationships and elements involved.

Consequently, as individual chapters will show, and as our conclusions in chapter 12 will discuss further, in respect of the reforms undertaken, there is considerable 'distance' between policy and impact. In particular, as federal research and innovation policies shift emphasis from inputs to outputs and outcomes there emerge intended and unexpected consequences in both the short term and longer term. In addition, changing federal and university research priorities are reshaping relationships among the various stakeholders. As we shall see, at the centre of these changes are a number of fundamental values and a range of issues and dilemmas.

Conclusions

This chapter has set out the overall objectives of the book, key definitional issues, and a brief initial view of the evolution of federal research and innovation policies in the Chrétien-Martin Liberal and Harper Conservative government periods.

It has also set out our overall analytical framework for analysing federal research and innovation policies and key changes in the relationships between the federal government and universities. The four-part classification framework has highlighted the need to examine: 1) high-level policy and conceptual discourse; 2) core policy values and ideas; 3) the main policy instruments and mixes thereof (taxation, spending, regulation, and persuasion); and 4) institutional and governance changes.

The chapter has established a starting point for a more integrated look at the more particular policy issues, institutions, relationships, and impacts which our authors examine, as well as some of the policy conflicts and value-based dilemmas inherently involved in what is ultimately a complex policy and institutional journey. In our conclusions in chapter 12 the editors bring together our views of the contributing authors' assessments of the impacts of federal research and innovation policies on universities and our own overall views about why and how federal research and innovation policies have changed the relationships between the federal government and universities in Canada and the dilemmas inherent in these changed relationships.

REFERENCES

AUCC. (2005). *Momentum: The 2005 Report on University Research and Knowledge Transfer.* Ottawa: AUCC. Available from: http://www.aucc.ca/momentum/en/report/ (accessed 30 November 2006).

Atkinson-Grosjean, Janet. (2006). *Public Science, Private Interests: Culture and Commerce in in Canada's Networks of Centres of Excellence.* Toronto: University of Toronto Press.

Aucoin, Peter. (1997). *The New Public Management: Canada in Comparative Perspective.* Montreal: McGill-Queen's University Press.

– (2003). 'Independent Foundations, Public Money and Public Accountability: Whither Ministerial Responsibility as Democratic Governance.' *Canadian Public Administration* 46 (1): 1–26.

Beach, Charles, M., Robin Boadway, and R.M. McInnis, eds. (2005). *Higher Education in Canada.* Montreal: McGill-Queen's University Press.

Boden, R.D., Cox, D., M. Nedeva, and K. Barker. (2004). *Scrutinising Science: The Changing UK Government of Science.* London: Palgrave Macmillan.

Bok, Derek. (2003). *Universities in the Marketplace: The Commercialization of Higher Education.* Princeton, N.J.: Princeton University Press.

Burke, Joseph C. (2005). *Quality and Performance in Higher Education: Baldrige on Campus.* New York: Anker Publishing.

Cameron, David M. (2003). 'The Challenge of Change: Canadian Universities in the Twenty-First Century.' *Canadian Public Administration* 45 (2): 145–74.

– (2004). 'Collaborative Federalism and Post-secondary Education: Be Careful What You Wish For.' Paper prepared for the John Deutsch Institute for the Study of Economic Policy, Queen's University, Kingston, Ontario.

– (2005a). 'Post Secondary Education and Research: Whither Canadian Federalism?' In *Taking Public Universities Seriously,* ed. F. Iacobucci and Carolyn Tuohy, 277–92. Toronto: University of Toronto Press.

– (2005b). 'Ontario's Rae Report: Investing in Growth.' *Canadian Public Administration* 48, no. 2: 280–7.

Canada. (1999a). *Public Investments in University Research: Reaping the Benefits.* Ottawa: Industry Canada.
- (1999b). *A New Approach to Health Research for the 21st Century: The Canadian Institutes of Health Research.* Ottawa: Canadian Institutes of Health Research.
- (2002). *Framework of Agreed Principles on Federally Funded University Research: An Agreement between the Government of Canada and the Association of Universities and Colleges of Canada.* Ottawa: Industry Canada.
- (2003). *Federal Science and Technology: The Pursuit of Excellence: A Report on Federal Science and Technology – 2003.* Ottawa: Industry Canada.
- (2005). *Canadian Institutes of Health Research 2005–2006: Report on Plans and Priorities.* Ottawa: Treasury Board Secretariat.
- (2006). *People and Excellence: The Heart of Successful Commercialization.* Ottawa: Industry Canada.
- (2007). *Mobilizing Science and Technology to Canada's Advantage.* Ottawa: Industry Canada.

Canadian Biotechnology Advisory Committee. (2006). *Human Genetic Materials, Intellectual Property and the Health Sector.* Ottawa: Canadian Biotechnology Advisory Committee.

Canadian Council on Learning. (2006). *Canadian Post-Secondary Education: A Positive Record – An Uncertain Future.* Ottawa: Canadian Council on Learning.

Clark, Burton R. (1998). *Creating Entrepreneurial Universities: Organizational Pathways of Transformation.* New York: Springer.

Crocker, Robert, and Alex Usher. (2006). *Innovation and Differentiation in Canada's Post-Secondary Institutions.* Ottawa: Canadian Policy Research Network.

Crompton, Helen. (2007). 'Mode 2 Knowledge Production: Evidence From Orphan Drug Networks.' *Science and Public Policy* 34 (3): 199–212.

Davenport, Sally, Shirley Leitch, and Arie Rip. (2003). 'The "Use" in Research Funding Negotiation Processes.' *Science and Public Policy* 30 (4): 239–50.

de la Mothe, John. (2003). 'Ottawa's Imaginary Innovation Strategy: Progress or Drift?' In *How Ottawa Spends 2003–2004: Regime Change and Policy Shift*, ed. G. Bruce Doern, 172–86. Don Mills, Ont.: Oxford University Press.

Delanty, G. (2001). *Challenging Knowledge: The University in the Knowledge Society.* Buckingham: Open University Press.

Doern, G. Bruce. (1972). *Science and Politics in Canada.* Montreal: McGill-Queen's University Press.
- (1995). *Institutional Aspects of R&D Tax Incentives: The SR&ED Tax Credit.* Ottawa: Industry Canada.

Doern, G. Bruce, ed. (2006). *Innovation, Science and Environment, 2006–2007: Canadian Policies and Performance.* Chapter 1, 3–34. Montreal: McGill-Queens University Press.

Doern, G. Bruce, ed. (2007). *Innovation, Science and Environment, 2007–2008: Policies and Performance.* Chapter 1, 3–31. Montreal: McGill-Queens University Press.

Doern, G. Bruce, and Jeff Kinder. (2007). *Strategic Science in the Public Interest: Canada's Government Laboratories and Science-Based Agencies.* Toronto: University of Toronto Press.

Doern, G. Bruce, and Richard Levesque. (2002). *NRC in the Innovation Policy Era: Changing Hierarchies, Networks and Markets.* Toronto: University of Toronto Press.

Doern, G. Bruce, and Richard Phidd. (1992). *Canadian Public Policy: Ideas, Structure, Process.* 2nd ed. Toronto: Methuen.

Doern, G. Bruce, and Markus Sharaput. (2000). *Canadian Intellectual Property: The Politics of Innovating Institutions and Interests.* Toronto: University of Toronto Press.

Doern, G. Bruce, Margaret Hill, Michael Prince, and Richard Schultz, eds. (1998). *Changing the Rules: Canadian Regulatory Regimes and Institutions.* Toronto: University of Toronto Press.

Doern, G. Bruce, and Robert Johnson, eds. (2006). *Rules, Rules, Rules, Rules: Multi-level Regulatory Governance.* Toronto: University of Toronto Press.

Dominion Bureau of Statistics. (1970). *Federal Government Expenditures on Scientific Activities.* Ottawa: Dominion Bureau of Statistics.

Doutriaux, Jerome. (2003). 'University-Industry Linkages and the Development of Knowledge Clusters in Canada.' *Local Economy* 18 (1): 63–79.

Etzkowitz, Henry, and Andrew Webster. (1998). 'Entrepreneurial Science: The Second Academic Revolution.' In *Capitalizing Knowledge: New Intersections in Industry and Academia*, ed. Henry Etzkowitz, Andrew Webster, and Peter Healey, 88–104. New York: SUNY Press.

Etzkowitz, Henry, and Loet Leydesdorff, eds. (1997). *Universities and the Global Knowledge Economy.* London: Pinter.

Fisher, Donald, et al. (2005). *Canadian Federal Policy and Post-Secondary Education.* Toronto: Alliance for International Higher Education Policy Studies.

Flinders, Matthew. (2002). *The Politics of Accountability in the Modern State.* London: Ashgate.

Geiger, Roger L. (2004). *Knowledge and Money: Research Universities and the Paradox of the Marketplace.* Palo Alto, Calif.: Stanford University Press.

Gibbons, Michael. (2000). 'Mode 2 Society and the Emergence of Context-Sensitive Science.' *Science and Public Policy* 27 (3): 33–41.

Gibbons, Michael, et al. (1994). *The New Production of Knowledge: The Dynamics of Science and Research in Contemporary Societies*. London: Sage Publications.

Government of Canada. (1973). *Guidelines for the Implementation of the Make Or Buy Policy Concerning Research and Development Requirements in the Natural Sciences*. Ottawa: Treasury Board Secretariat, Administrative Policy Branch. January.

– (1975). *The Make-or-Buy Policy, 1973–1975*. Ottawa: Ministry of State for Science and Technology. November.

– (1977). *Policy and Guidelines on Contracting Out the Government's Requirements in Science and Technology*. Ottawa: Treasury Board Secretariat, Administrative Policy Branch.

Guston, David. (2000). *Between Politics and Science*. Cambridge and New York: Cambridge University Press.

Health Canada. (2007). *Human Genetics Licensing Symposium*. Ottawa: Health Canada.

Horn, Michiel. (1998). *Academic Freedom in Canada: A History*. Toronto: University of Toronto Press.

Howlett, Michael, and M. Ramesh. (2003). *Studying Public Policy*. 2nd ed. Don Mills, Ont.: Oxford University Press.

Iacobucci, Frank, and Carolyn Tuohy, eds. (2005). *Taking Public Universities Seriously* Toronto: University of Toronto Press.

Ibbitson, John. (2009). 'U.S.-Canada Divide is Stark on Science.' *Globe and Mail*, 27 April, 7.

Jones, Glen A. (2005). 'On Complex Intersections' Ontario Universities and Governments.' In *Taking Public Universities Seriously*, ed. Frank Iacobucci and Carolyn Tuohy, 174–87. Toronto: University of Toronto Press.

Jones, Glen, P. McCarney, and M. Skolnik, eds. (2005). *Creating Knowledge, Strengthening Nations: The Changing Role of Higher Education*. Toronto: University of Toronto Press.

Kinder, Jeff. (2003). 'The Doubling of Government Science and Canada's Innovation Strategy.' In *How Ottawa Spends 2003–2004: Regime Change and Policy Shift*, ed. G. Bruce Doern, 204–20. Don Mills, Ont.: Oxford University Press.

Laidler, David, ed. (2002). *Renovating the Ivory Tower: Canadian Universities and the Knowledge Economy*. Toronto: C.D. Howe Institute.

LaPointe, Russell. (2006). 'The Social Sciences and Humanities Research Council: From a Granting Council to a Knowledge Council.' In *Innovation, Science, Environment, 2006–2007: Canadian Policies and Performance*, ed. G. Bruce Doern, 127–48. Montreal: McGill-Queen's University Press.

Lopreite, Débora. (2006). 'The Natural Sciences and Engineering Research Council as a Granting and Competitiveness-Innovation Body.' In *Innovation,*

Science, Environment 2006-2007: Canadian Policies and Performance, ed. G. Bruce Doern, 105–26. Montreal: McGill-Queen's University Press.

Lundvall, Bengt-Åke. (2004). *Why the New Economy is a Learning Economy.* DRUID Working Paper No. 2004–01. Copenhagen.

Luukkonen, Terttu, Maria Nedeva, and Remi Barre. (2006). 'Understanding the Dynamics of Networks of Excellence.' *Science and Public Policy* 23 (4): 239–52.

Marginson, Simon, and Mark Considine. (2000). *The Enterprise University: Power, Governance and Reinvention in Australia.* Melbourne: Cambridge University Press.

Murphy, Joan. (2007). 'Transforming Health Sciences Research: From the Medical Research Council to the Canadian Institutes of Health Research.' In *Innovation, Science, Environment: Canadian Policies and Performance, 2007–2008,* ed. G. Bruce Doern, 240–61. Montreal: McGill-Queen's University Press.

Nelles, Jen, Allison Bramwell, and David A. Wolfe. (2005). 'History, Culture and Path Dependency: Origins of the Waterloo ICT Cluster.' In *Global Networks and Local Linkages: The Paradox of Cluster Development in an Open Economy,* ed. David A. Wolfe and Matthew Lucas, 110–24. Montreal and Kingston: McGill-Queen's University Press for the School of Policy Studies, Queen's University.

Nelson, Richard R., and Sidney G. Winter. (1982). *An Evolutionary Theory of Economic Change.* Cambridge, Mass.: Belknap Press.

Nelson, Richard R., ed. (1993). *National Innovation Systems: A Comparative Analysis.* Oxford and New York: Oxford University Press.

OECD. (1999). *Cluster Analysis and Cluster-Based Policy in OECD Countries.* Paris: Organization for Economic Co-operation and Development.

Pal, Leslie A. (2006). *Beyond Policy Analysis: Public Issue Management in Turbulent Times.* 3rd ed. Toronto: Nelson.

Pavitt, Keith. (1991). 'What Makes Basic Research Economically Useful?' *Research Policy* 20 (2): 109–19.

Pocklington, T., and A. Tupper. (2002). *No Place to Learn. Why Universities Aren't Working.* Vancouver: University of British Columbia Press.

Polanyi, Michael. (1967). *The Tacit Dimension.* Garden City, N.Y.: Doubleday Anchor Books.

Polster, C. (2003–4). 'Canadian University Research Policy at the Turn of the Century: Continuity and Change in the Social Relations of Academic Research.' *Studies in Political Economy* 71/72: 177–99.

– (2007). 'The Nature and Implications of the Growing Importance of Research Grants to Canadian Universities and Academics.' *Higher Education* 53 (5) (May): 599–622.

Rosenblatt, Michael, and Jeffrey Kinder. (2006). 'Canadian and UK Innovation Policies: A Comparative Analysis,' In *Innovation, Science and Environment: Canadian Policies and Performance, 2006–2007*, ed. G. Bruce Doern, 59–81. Montreal: McGill-Queen's University Press.

Schillo, Sandra. (2003). *Knowledge Transfer and Commercialization in Canada's Federal Laboratories*. Ottawa: Advisory Council on Science and Technology.

Science, Technology and Innovation Council. (2009). *State of the Nation 2008: Canada's Science, Technology and Innovation System*. Ottawa: Science, Technology and Innovation Council.

Shove, Elizabeth, and Arie Rip. (2003). 'Users and Unicorns: A Discussion of Mythical Beasts in Interactive Science.' *Science and Public Policy* 27 (3): 175–82.

Slaughter, Sheila, and Gary Rhoades. (2004). *Academic Capitalism and the New Economy*. Baltimore and London: The Johns Hopkins University Press.

Statistics Canada. (2006). *Federal Government Expenditures on Scientific Activities, 2006/2007*. Statistics Canada, Catalogue 88-001-XIE, vol. 30, no. 6 (September).

Swimmer, Gene, ed. (1996). *How Ottawa Spends, 1996–97: Life Under the Knife*. Chapter 1, 1–38. Ottawa: Carleton University Press.

Tamburri, Rosanna. (2009). 'Schools May Curtail Hiring, Scholarships.' *University Affairs* (January): 25–7.

Tupper, Allan. (2003). 'The Chrétien Governments and Higher Education Policy: A Quiet Revolution in Canadian Public Policy.' In *How Ottawa Spends, 2003–2004: Regime Change and Policy Shift*, ed. G. Bruce Doern, 105–17. Don Mills, Ont.: Oxford University Press.

Wolfe, David, ed. (2003). *Clusters Old and New: The Transition to a Knowledge Economy in Canada's Regions*. Kingston: School of Policy Studies, Queen's University.

Wolfe, David A., and Matthew Lucas, eds. (2004). *Clusters in a Cold Climate: Innovation Dynamics in a Diverse Economy*. Montreal and Kingston: McGill-Queen's University Press for the School of Policy Studies, Queen's University.

2 Pushing Federalism to the Limit: Post-Secondary Education Policy in the Millennium

ALLAN TUPPER

This book's focus on research and innovation policy and the changing relations between universities and the federal government must first be seen in the context of the larger domain of federalism and post-secondary education (PSE) policy. Post-secondary education is a policy area in Canada with many interesting characteristics and controversies that play out usually amongst specialized constituencies rather than at a mass political level. First, PSE is a substantial expenditure realm for both the federal and the provincial governments. In every province, it is the third largest expenditure behind only health and 'K to Twelve' education. In 2006–07, for example, the governments of Alberta, British Columbia, and Ontario budgeted $2.2 billion, $2 billion, and $3.9 billion respectively for their ministries of advanced education. Second, post-secondary education is now seen as a major policy area with important implications for many other major areas of government activity. As Keith Banting (2006) argues, access to education is, in a competitive, interdependent world where traditional income support measures are considerably weakened, *the* major source of economic security for citizens. Third, no consensus has emerged in Canada and other countries about the ultimate goals of higher education policy. How closely should education be linked to labour markets? Can the private sector play a major role in higher education? Are broad access and high-quality post-secondary education compatible?

A small circle of government policymakers, student leaders, and institutional leaders debate these questions with intermittent interventions from business leaders and the media. Typically, post-secondary education policy does not engage mass publics even as many more Canadians depend on, and attend, universities and colleges and even though post-secondary education policy can evoke strong differences of opinion with regard to access, funding, and goals. Finally, Canadian federalism, the precise subject of this chapter, deeply shapes the substance and processes of higher-education policymaking. Both federal and provincial governments have major roles but operate without formal coordination. Such intergovernmental relations are labelled 'uncoordinated entanglement.'

Canada's post-secondary education 'system' is admittedly imperfect.[1] But in the larger scheme of things, Canada has an accomplished, high-quality post-secondary education system. Federalism has both permitted and contributed to that outcome. On the other hand, Canadian federalism, an admittedly adaptive institution, is being tested by post-secondary education. Three complex forces confront it: 1) the complexity of post-secondary education issues in the new economy; 2) the ambitions of universities; and 3) Canadians' need for greater access to post-secondary education.

The chapter first examines federal and provincial roles in post-secondary education. It pays particular attention to the period from the mid-1990s to the present and highlights Ottawa's capacity to act decisively in post-secondary education, with major consequences for universities, students, and provincial government policy. As chapter 1 has indicated, Ottawa's substantial expansion of its support for university research since 1997 illustrates the uneasy co-existence of a major provincial government role in institutional governance, operating funding, and student financial assistance, with a dominant federal presence in research funding. In a nutshell, Ottawa's preoccupation is advanced research policy, while the provinces deal with the instructional roles of universities and colleges.

The chapter's second section focuses on the Conservative minority government of Stephen Harper. Two points are addressed. First, do the Conservatives, like their Liberal predecessors, see post-secondary education as a major priority? Second, what are the consequences of post-secondary education's increasing importance in federal–provincial relations? A 'dedicated intergovernmental transfer' for higher education is widely discussed in the context of changes to intergovernmental finances. Post-secondary education is now of considerable interest to

---

1 The word 'system' is used here very loosely. Neither Canada nor some provinces have systems of post-secondary education as understood in many other countries and jurisdictions. Hence the term 'uncoordinated entanglement.' See Teichler 2007.

the Council of the Federation, whose mandate and grasp elevates it to the apex of interprovincial relations. Whether post-secondary education policy will be well served by its possible entry into 'summit federalism' is speculated on.

The chapter's third section explores the relationship between Canadian federalism and post-secondary education policy. It argues that Canadian federalism as currently structured precludes a necessary overhaul of student financial assistance and the emergence of a more explicitly differentiated university system, where a few large universities stress research and graduate studies and others emphasize undergraduate teaching. Finally, it speculates about the future of post-secondary education and Canadian federalism. Three conclusions emerge. First, changes are required if policy is to be more effective. The federal and provincial governments must better coordinate their post-secondary education policies. Second, universities themselves will also have to change. They will have to be less concerned with their explicitly economic roles and more concerned with effective undergraduate education, community service, and broad accessibility. Third, federalism is itself an obstacle to reform. Neither level of government shows a serious inclination to build new relationships in post-secondary education. For different reasons, Ottawa and the provincial governments have interests in the status quo. Hence the ultimate adaptability of federalism is an open question. As Jennifer Smith reminds us, federalism is always 'in flux and in cement' (2004, 8).

Post-Secondary Education and Canadian Federalism: Uncoordinated Entanglement

Canadian governments are deeply involved in post-secondary education. The common assertion, that Ottawa funds research and the provinces fund teaching, is both correct and a bit misleading at the same time. It also implies a non-existent tidiness in policy. The Government of Canada is extensively involved in the funding of research. But, as chapter 1 has shown, in the era of the Canada Foundation for Innovation, research embraces substantial funding for research infrastructure, a matter of great interest to provincial governments. Ottawa, through the Canada Millennium Scholarship Foundation, the income tax system, and the Canada Student Loans Program, is involved in student financial assistance, an area that preoccupies provincial governments and one that is quite different from the strong federal focus on research.[2] And since its 2002 decision to fund partially the 'indirect costs' of federally funded university research, the federal government provides something like an operating grant to universities. Support for post-secondary education was originally a rationale for equalization and the modern regime of federal unconditional grants to provincial governments. In summary, Ottawa directly funds post-secondary institutions, students, and provincial governments.

Formal constitutional authority over education rests with provincial governments, with the caveat that a system of mass public education, let alone research universities, was not envisioned in 1867. At present, Canadian provincial governments preside over tuition fees, student numbers, the structure of provincial higher education systems, institutional governance, and labour relations. They create, disband, and reorganize institutions. They approve degree programs, assess their quality, and provide substantial operating and capital funds to universities. Each province has programs of student financial assistance that include loans, bursaries, scholarships, and income tax measures.

Ottawa dominates research funding but provincial governments, especially Ontario, Quebec, and Alberta, are also active. They endow research chairs, encourage research in important areas, and directly fund major research programs. Alberta's Heritage Foundation for Medical Research, for example, has a substantial endowment, now in excess of $1.2 billion, whose resources have been used to build major health research complexes in Edmonton and Calgary.

In the late 1950s and early 1960s, the provincial governments undertook a substantial expansion of higher education driven by post-war affluence, the entry of large numbers of women, and the coming to age of baby boomers. They built universities and colleges and promoted graduate education. Student financial assistance was, by contemporary standards, generously provided, often through substantial grants.

The heady days of 1960s province-building were replaced by less ambitious provincial policies in the 1970s and early 1980s. In the late 1980s and through the 1990s, the provinces adopted aggressive post-secondary education policies that were often unpopular with institutions and with students. Driven by provincial budget deficits and growing debts, conservative political ideas and federal restraint,

---

2 The 2008 Federal Budget noted that the Canada Millennium Scholarship Foundation would not be renewed. As a result, the Foundation's ability to make awards expires in January 2010 as envisioned in its 1998 legislation.

provincial governments, in different ways and at different times, took a hard line on post-secondary education. Operating grants were often meagrely increased. Provincial governments became more directive and shaped university and college priorities with specialized programs that reflected particular needs. Universities and colleges were subjected to new reporting and accountability processes that they saw as intrusive. In several provinces faculty and staff salaries were subject to provincial controls, and in Alberta salaries were cut in the early 1990s. Tuition rose steadily except in Quebec, but in distinctive ways in each province. Ontario, the most populous province, was a particular champion of stern post-secondary education policies. Ian Clark, president of the Council of Ontario Universities, estimated in 2003 that Ontario universities received only 52 per cent of their operating funds from Queen's Park, compared to slightly more than 75 per cent a decade earlier (Clark 2003, 413).

The structure of Canada's post-secondary education system reflects the considerable diversity of the Canadian provinces, each of which has developed a relatively unique post-secondary education system. For example, universities, colleges, and professors have national associations that advance common causes and build national consciousness. Equally, such organizations have provincial counterparts that perform comparable activities at the provincial level, generally without explicit coordination with national associations. Second, provincial post-secondary education systems vary considerably in their structure and, to a lesser degree, roles. For example, colleges, universities, and technical institutes have different relationships with each other and with their provincial governments. Tuition policy, admissions criteria, and student financial assistance vary considerably in detail, although only Quebec has steadfastly maintained very low tuition fees for provincial residents. Third, as in other countries/jurisdictions, no national system exists for credit transfer either between post-secondary institutions or between secondary and post-secondary education systems in the provinces. No federal-provincial body has a clear mandate to ensure national standards, to encourage student mobility, or to think carefully about system capacity, access, or student numbers.

Impressive credentials underpin a major role for the Government of Canada in post-secondary education. The Royal Commission on National Development in the Arts, Letters, and Sciences, the Massey Commission, spoke passionately about the nation-building roles of universities and the importance of accessible post-secondary education

(Canada 1951). In 1958, the Royal Commission on Canada's Economic Prospects (Canada 1958), the Gordon Commission, addressed, with contemporary resonance, Canada's urgent need to expand universities in the face of labour shortages and a changing continental economy that demanded more educated workers. It urged Ottawa to better fund research and graduate studies and to upgrade university physical plant. As the Commission put it: 'We do, however, feel it is our bounden duty to call attention as forcefully as we can to the vital part which the universities must play in our expanding and increasingly complex economy, and to the necessity of maintaining them in a healthy and vigorous condition. The functions of universities touch every facet of our society' (Canada 1958, 452). The Royal Commission on the Economic Union and Development Prospects for Canada, the Macdonald Commission, studied Canada's economy in the 1980s (Canada 1985). It adopted a sceptical, somewhat technocratic approach to post-secondary education. The Macdonald Commission acknowledged, however, that Ottawa had a major role to play. Given the mobility of well-educated persons, provincial governments had an incentive to underfund higher education, lest graduates leave the province. As a result, Ottawa had a clear rationale for funding higher education.

The Government of Canada has sometimes moved decisively and unilaterally in post-secondary education policy (Cameron 2003; 2004; 2005a). After the Second World War, Ottawa drove a major expansion of Canadian universities by providing tuition and living expenses to returning veterans who confronted a changing economy. In addition to direct support for veterans, Ottawa also provided universities with financial support to accommodate the massive influx. In 1951, it adopted the Massey Commission's recommendation that Ottawa provide unconditional grants directly to universities. Federalism made its mark on this national initiative when Quebec initially accepted and then refused federal funding. It maintained its opposition until 1959, when Ottawa agreed to provide funds directly to the Government of Quebec.

In 1966, Ottawa implemented a radical policy change when it replaced grants to universities with unconditional grants to provincial governments. As David M. Cameron notes: 'Universities, at least in English Canada, could hardly believe what had happened. Henceforth, they would have to rely on provincial operating funding decisions alone for the general operating support (along with tuition revenue) and they would soon discover that provincial funding would fall short of their needs' (Cameron 2004, 11). In 1995, Ottawa again acted aggressively, with poor results for post-secondary education when it rolled the Established Programs Financing and the Canada Assistance Plan into a single transfer – the Canada Health and Social Transfer (CHST) – that covered health, post-secondary education, and social assistance. More painfully, CHST cut federal funding for these major public priorities by $6 billion, adding momentum to the pattern of declining provincial operating support for post-secondary institutions and rising tuitions. In 2005, Ontario provided a stark example of the cumulative impact of this pattern. In several Ontario universities, tuition revenues were roughly on par with provincial operating grants as contributors to total university revenues (Council of Ontario Universities 2006).

The Chrétien Governments: Universities at Centre Stage

In 1997, Ottawa embarked on a new course in post-secondary education. Backed by a budget surplus, it launched substantial initiatives that had transformative results for Canadian universities. As chapter 1 has shown, the federal government began to see universities as important sites for innovation in a changing economy. Observers are unanimous that 'a new paradigm' was envisioned whereby universities would work closely together, research capacity would be strengthened substantially, and more university research would be 'commercialized' (Bakvis 2007; Prichard 2000; Tupper 2003). In Ottawa's mind, strong universities and high-quality research, if effectively harnessed to national priorities, would boost Canada's economic competitiveness. The education of 'highly qualified personnel,' a key element of a modern economy, was a particular university obligation.

Universities, especially their senior leadership, strongly supported Ottawa's aggressive renewal of university research. As chapter 3 in this book shows, the Association of Universities and Colleges of Canada (AUCC) skilfully linked universities' priorities with federal government needs and campaigned vigorously for increased funding. Its work was reinforced by several politically astute presidents of major research universities, who built close relationships with senior officials in Finance Canada, Industry Canada, and the Prime Minister's Office (Bakvis 2007; Tupper 2003). Canadian universities wanted a research environment comparable to that prevailing in the best American research universities. The major elements of a 'world class' research university, and implicitly a dynamic economy, were said to be outstanding research faculty, numerous graduate students, excellent infrastructure, and substantial funding for the direct and indirect costs of research. Ottawa undertook programs in all these areas.

As chapter 5 shows, Ottawa's first major initiative was the Canada Foundation for Innovation (CFI), whose principal role was to invigorate research infrastructure at Canadian universities. Chapter 5 indicates several ways in which the CFI has been criticized, but also that such criticism has not lessened university support for the CFI. It is too valuable a source of funding for necessary equipment and infrastructure.

In 1999, a second major initiative, the Canada Research Chairs program, dealt with outstanding research professors, a natural complement to CFI's focus on infrastructure. The CRC program provided $900 million to fund two thousand research professors, half of whom were to be senior, proven performers (Tier I CRCs), while the other half were promising younger faculty members (Tier II CRCs). This substantial program allowed Canadian universities to recruit and retain outstanding researchers in all fields. It generated professorial mobility (and higher salaries) both within Canada and from other countries to Canada. Some Canadians were lured back from the United States and overseas while outstanding non-Canadians were also recruited.

Perhaps Ottawa's most important recent contribution was the introduction of a program of support for the 'indirect costs' of federally funded research. Unlike the United States and other major countries, the Government of Canada for many years funded only the direct costs of university research. Over time, universities shouldered the 'indirect costs,' the overhead associated with research grants. By conservative estimates, universities concluded that indirect costs were at least 40 cents on each federal research dollar. This situation, which had been complained about for decades by Canadian universities, worsened as the 'new paradigm' put more money into university research. The indirect cost problem was particularly acute for universities with major research capacity, especially in high-cost areas. As chapter 1 has suggested, recent federal overhead payments would barely cover 12 to 14 per cent of such costs (see also the discussion in chapter 4 on the granting councils and in chapter 5 on the Canada Foundation for Innovation).

Put simply, the more successful the university in receiving federal research grants, the greater its indirect costs 'deficit.' Students, notably the undergraduate majority, paid the deficit through higher tuition, deteriorating infrastructure, and ultimately, a lesser educational experience as faculty success in obtaining federal research grants drained university resources. A one-time federal allocation of $200 million was provided in 2001, followed by a permanent program funded at $260 million in 2006.

Vigorous federal leadership in university research was complemented by measures for students. In 2003, the Canada Graduate Scholarships, presently funded at $105 million, were established to encourage Canadian students to pursue graduate studies at Canadian universities. A more substantial undertaking was the Canada Millennium Scholarships Foundation, established in 1998 with a $ 2.5 billion endowment and a mandate to provide bursaries to large numbers of students. Hastily conceived, the Millennium Foundation created serious conflict with provincial financial assistance programs. Its political impact, however, when combined with federal income tax changes like the Registered Educational Savings Plan and the Canada Education Savings Grant, was to make the Government of Canada appear as a benign presence in student (and parental) financial assistance. As provinces raised tuition and eliminated grants, Ottawa announced programs that seemed, or claimed, to increase access and affordability.

This substantial infusion of $11 billion in federal funds, whose major components are noted in Table 2.1, has had major positive impacts on Canadian universities (Canada 2006a). CFI has substantially strengthened university research infrastructure and provided long overdue modernization. The CRC program, a major undertaking by any country's standard, has had benign impacts on universities. Federal support for the indirect costs of research responded to a perennial university concern. In recent years, no other major country has moved as decisively as Canada with new funding and programs for university research.

Ottawa's 'new paradigm' caused spirited debate on Canadian university campuses, although university leadership has strongly supported federal policy (Prichard 2000). Some observers worry that research and federal government priorities now exert excessive influence on university policymaking. Others argue that smaller universities have not gained their fair share and that program criteria work against their interests. Perennial debates have been rekindled about the balance between social sciences and humanities, natural sciences and health sciences. Some critics assert that Ottawa has been too timid and traded off research excellence for equitable regional distribution of funding. A variation of this debate is whether research funding should be concentrated in a few outstanding research universities.

Table 2.1
Government of Canada: Major Programs for Support of University Research and Students, 1997 to 2006*

Organization or program	Date of establishment	Initial Proposed financing/ endowment	Major purpose
Canada Foundation for Innovation	1997	$800 million endowment	Modernize and strengthen university research infrastructure
Canada Millennium Scholarship Foundation	1998	$2.5 billion endowment	Provide bursaries to Canadian students
Canada Research Chairs	1999	$900 million	Recruit and retain outstanding university researchers
Canadian Institutes of Health Research (previously Medical Research Council of Canada)	2000	$464 million ($289 million base from predecessor organization, the Medical Research Council and incremental funding of $175 million)	Promote new approaches to health research, including greater emphasis on research, commercialization, and interdisciplinary work
Indirect Costs of Federally Funded University Research	2001	$200 million, non-recurring	Provides funding for research overheads directly to institutions
Canada Graduate Scholarships	2003	$105 million	Support for outstanding Masters and Doctoral Students for study at Canadian universities
Post-Secondary Education Infrastructure Trust	2006	$1 billion, non-recurring	Support for provincial government investments in teaching facilities and equipment

* Excludes transfers to provincial governments for post-secondary education under the Canada Social Transfer (and its predecessors) and 'tax expenditures' for post-secondary education.

Aggressive federal leadership in university research has had major consequences for federal–provincial relations. It highlights the 'uncoordinated entanglement' of post-secondary education policy. Ottawa's recent post-secondary education initiatives are justified by federal spending power and involve direct federal relationships with post-secondary institutions, researchers, and students. None of the programs is a federal-provincial initiative in design, although all recent federal initiatives have major impacts on provincial policy and public finances.

As chapter 5 shows in detail, the CFI substantially refurbished university research infrastructure. Ultimately, the provincial governments have been CFI's main funding partners. In several provinces, new administrative structures have been established to provide what universities tellingly refer to as 'matching funds.' Recall, however, that CFI projects are assessed according to federal government priorities in refereed national competitions, without direct provincial government input. Would provinces have funded the same research infrastructure or research infrastructure at all? Would CFI 'matching funds' be deployed differently under a different set of program incentives? Are provincial governments (and universities) certain that longer-term operating costs have been carefully considered?

The establishment of the Millennium Foundation was particularly trying for provincial governments. As David M. Cameron (2004) has noted, the Millennium Foundation was hastily created, without careful consideration of its impact on provincial financial assistance plans. Ultimately, a decision was made to deliver Millennium funds through existing provincial government machinery. As a result, Millennium bursaries have been integrated in different ways in different provinces. At the end of the day, they may simply have displaced existing provincial funding. Moreover, the Millennium Foundation is scheduled to end operations in 2010 with the result that considerable federal funds will be withdrawn from the student financial assistance pool. More generally, federal support for university research and its indirect costs are components of the annual expenditure budget where, like all such public expenditures, they are at the mercy of broader political and economic trends and priorities. What happens if they are cut substantially?

Ottawa's new research paradigm has heightened the consciousness of universities as national institutions closely linked to the federal government. It has correspondingly reduced their interest in provincial governments, whose interests seem parochial when compared to the dynamic national scene. Universities have received a considerable portion of the federal budget surpluses for more than a decade in the face of competing priorities. As political actors, they have learned the need for common fronts if success is to be achieved in federal budget politics. They have become a powerful national force, likely with the unanticipated consequence that they are, or are seen to be, 'simply another pressure group.'

The 'new paradigm' established federal institutions and administrative processes that reinforced the importance of university research and linked university leaders with senior public servants and business leaders. A striking symbol of the deep bonds between universities and the Government of Canada is the 2002 'Framework of Agreed Principles on Federally Funded University Research' (Canada 2002). It commits universities, subject to adequate federal funding, to ambitious, quantifiable increases in research activities, including 'commercialized research.'

The Harper Conservatives and Post-Secondary Education: A New Page? A New Chapter?

In November 2005, Paul Martin's Liberal government released a *Fiscal and Economic Update* (Canada 2005) that noted the strength of the Canadian economy and the healthy state of federal finances. It predicted large budget surpluses for the foreseeable future and promised even more financial support for post-secondary education. Among other things, large increases were promised for Canada Graduate Scholarships, the Canada Foundation for Innovation, indirect costs of research, and the federal research granting councils. For universities, the Promised Land looked even better.

The 2006 general election resulted in a Conservative minority government led by Stephen Harper. Canadian post-secondary education institutions were apprehensive. They were accustomed to federal Liberal governments that spoke their language and funded them well. The Harper Conservatives, on the other hand, were little known especially to post-secondary education elites in Ontario and Quebec. Their convictions and ambitions were neither well known nor understood.

Post-secondary education issues were noted but were certainly not central to Conservative campaigns in the general elections of 2004 and 2005–06. Harper himself had once advocated the construction of a 'firewall' around Alberta and a substantial reduction of federal government influence in that province. The Conservative campaign stressed accommodation with the provincial government of Jean Charest and implied, or was seen to imply, a general disposition toward provincial government power. It sometimes spoke of rectifying the 'fiscal imbalance,' a term used by provincial governments, especially Quebec, as a descriptor of their dissatisfaction with intergovernmental finances. The Harper Conservatives advocated a 'dedicated' federal transfer to the provinces for post-secondary education. Universities feared that the Conservatives might limit direct federal transfers to universities for research in favour of student financial assistance and/or unconditional transfers to the provincial governments.

At the time of writing neither the Conservatives' final stance on post-secondary education nor the government's political prospects are clear. That said, the 2006 and 2007 Harper government federal budgets were encouraging, although post-secondary education policy was not their focus. As chapter 1 has shown, they forecasted large surpluses, further funded the indirect costs program, and increased funding to the federal research granting councils. In 2006, students received a tax credit for textbooks. Scholarship and bursary income was made non-taxable. The Conservatives also committed to a $1 billion Post-Secondary Education Infrastructure Trust, subject to a 2005–06 surplus in excess of $2 billion. The fund was to be allocated to provincial governments on a per capita basis. As the budget put it: 'The Post-Secondary Education Infrastructure Trust will support critical and urgent investments that will enhance universities' and colleges' infrastructure and equipment (e.g. modernizing classrooms and laboratories; updating training equipment), as well as related institutional services (e.g. enhancing library and distance-learning technologies)' (Canada 2006a, 83). The Infrastructure Trust is an interesting federal recognition of the need for general teaching and learning infrastructure – areas related to, but quite different from, those funded by CFI.

Less positively, universities were worried in 2006 by rumours that the Treasury Board planned to cut funding to such important programs for international education as the Canadian Studies Program, the Commonwealth Scholarships, and the Canada–U.S. Fulbright Program. The government apparently saw these programs as primarily 'educational,' and hence properly undertaken by the provinces. Strong objections from the AUCC may have caused the government to reconsider its position. The threat of the elimination of programs, one not heard for many years by universities in Ottawa, caused apprehension notwithstanding otherwise positive budget signals.

The 2007 federal budget addressed post-secondary policy in several ways (AUCC 2007). The Canada Social Transfer, the major federal government financial transfer for post-secondary education policy, major social programs, and support for children, was extended until 2013–14. Ottawa established funding levels, within the CST, for each of the transfer's three main priorities. In this regard, the Conservatives may argue that a 'dedicated transfer' for post-secondary education has been delivered. Importantly, a 3 per cent annual escalator was included. A further total of $85 million was provided to the three major federal research granting councils, although such support was to be devoted to research deemed by the government to be important to Canada's national interest. Conservative research priorities included energy, information technologies, management, finance, and the environment. An additional $15 million was provided for indirect costs of research. Universities complained, as they had done to the Chrétien and Martin Liberal governments, that the indirect cost program was poorly funded. This remains largely true under the Conservatives. More money was provided for Canada Graduate Scholarships, whose various programs were named after famous Canadian researchers and innovators, including Sir Frederick Banting, Alexander Graham Bell, and Joseph-Armand Bombardier.

The Harper government appears to be serious about reforming intergovernmental finances, but likely needs a majority to implement fully its ideas. In a widely discussed supplement to the 2006 budget, *Restoring Fiscal Balance in Canada: Focusing on Priorities*, the government articulated its ideas about federal–provincial financial relations (Canada 2006b). Post-secondary education played a prominent role in this discussion. Major changes may occur.

The Conservatives' discussion of federalism gives no hints of a 'provincialist' or decentralizing philosophy. On the contrary, it asserts the spending power of and outlines a major role for the Government of Canada. The government's major concerns are well-known ones about program overlap, lack of predictability in intergovernmental finances, and excessive entanglement in some areas. *Restoring Fiscal Balance* notes that federal surpluses have frequently been used for spending power programs that have complicated intergovernmental relations. Three problem areas are noted: early childhood development, childcare, and housing and homelessness. Federal programs in these areas have created uncertainty for provincial governments with respect to the stability of funding and have often been introduced without consultation (Canada

2006b, 22–3). Interestingly, the major recent federal programs for university research are 'spending power' programs with similar intergovernmental consequences for those mentioned by the government. Those programs were mentioned positively and without reference to adverse consequences for intergovernmental relations.

Restoring Fiscal Balance in Canada noted Ottawa's major role in post-secondary education. 'The federal government is a long-standing contributor to post-secondary education (PSE) in Canada. The system of universities and colleges that exists today reflects decades of collaboration and cooperation between the two orders of government' (Canada 2006b, 35). The Conservatives then note the expansion of direct federal financial assistance to Canadian universities, colleges, and students over the past decade. The observation is made frequently that the 'mix' of federal support for higher education has altered over the last decade. Direct spending has waxed while transfers to the provinces have waned. 'Federal cash transfers for post-secondary education – an estimated $2 billion of the CST – have declined as a share of total post-secondary education expenditures' (Canada 2006a, 44). In turn, the combination of reduced federal cash transfers and reduced provincial operating funds caused tuition fees to rise dramatically in most provinces.

In short, the Harper Conservative government has increased support for granting councils and indirect costs, provided further, admittedly incremental, support for students, and asserted a major role for Ottawa in post-secondary education. Except for their controversial plan to eliminate international educational programs, the Conservatives have acted and spoken positively. One area that merits close attention, however, is the proposed changes to the Canada Social Transfer (CST). The 2007 federal budget suggested that incremental CST funding was dependent on federal–provincial agreement about the use of such funds by provincial governments. As Herman Bakvis observes, Quebec, and likely other provinces, might object in principle to this idea because it makes the CST less 'unconditional' (Bakvis 2007, 220).

The picture is more complex when provincial activities are examined. Since 2004, the provincial governments have shown an interest in working collectively through the Council of the Federation on post-secondary education where, to a degree, they are driven by a common desire to gain greater unconditional federal support. Two major activities highlight renewed provincial interest. First, the Council convened a Post-Secondary Education and Skills Summit in February 2006 in Ottawa co-chaired by Premiers Charest and McGuinty. The summit brought together 'stakeholders' selected by provincial governments to discuss major issues, including access to post-secondary education, funding, and research. The summit reached few specific conclusions but did send a clear message that the provincial governments remained major actors in post-secondary education.

A second important provincial initiative was the Council of the Federation's major study on the 'fiscal imbalance.' Its report, *Reconciling the Irreconcilable: Addressing Canada's Fiscal Imbalance*, asserted a large fiscal imbalance in favour of the Government of Canada. It argued that large federal surpluses were themselves evidence of the fiscal imbalance, as was vigorous federal use of the spending power. The Council's study saw three main avenues of federal influence over universities: unconditional transfers through the CST, direct transfers for research and to students, and the activities of foundations like CFI. While acknowledging the benign federal role in research, the Council report painted a bleak overall picture of Ottawa's influence on post-secondary education. It claimed that provincial post-secondary infrastructure was strained by federal research expenditures, and that provincial planning was impaired by lack of consultation and fears that Ottawa might abruptly terminate programs. It concluded: 'direct federal expenditures in post-secondary education often needlessly complicate policy-making and create a patchwork of programs and initiatives that waste resources and frustrate students, faculty and public servants' (Council of the Federation 2006, 60). This harsh assessment, whatever its merits as a piece of policy analysis, is certainly at variance with the views of Canadian universities in 2007. As a solution, the report recommended a dedicated transfer to start at $4.9 billion, with a 4.5 per cent annual escalator.

In 2007–08, post-secondary education garnered more attention as a major issue in Canadian federalism. The Conservative government, while not as rhetorically enthusiastic as its Liberal predecessors, is taking a benign view. The premiers have declared post-secondary education to be a priority that requires work on their part individually, through the Council of the Federation and with the Government of Canada.

Post-secondary education institutions and students should be alert, however, to the implications and potential drawbacks of their possible new importance in federal–provincial relations. Policy issues assume a different character when governments engage them at the highest levels. They become entangled in a web of complex policy issues. For example, the issue of a 'dedicated transfer' for post-secondary institutions is a small element of a larger debate about equalization, spending power, and health care. That debate in turn engages Quebec's concerns about a structural 'fiscal imbalance,' Ontario's fury about equalization, and Alberta's resource wealth. These matters will ultimately be addressed in federal–provincial meetings, perhaps at the first ministers' level. In 'summit federalism,' the status concerns of government, entrenched policy positions, and 'fiscal federalism' issues are the driving forces (Simeon and Robinson 2004). As noted, two Conservative changes to the CST – the specification of funding levels for each of its three main priorities and the idea that incremental funding will require some element of federal approval of provincial spending priorities – might evoke a provincial critique phrased as a matter of principle.

The high politics of Canadian federalism are the closed domain of an elite of financial experts, senior officials in central agencies, influential ministers, and government leaders. This small world is difficult to access and communicate with. It certainly hears 'interest group' demands, but often faintly. In short, post-secondary education 'stakeholders' should not necessarily rejoice in their ascent in the intergovernmental hierarchy. Their relatively stable, although not always harmonious, relations with provincial ministries of advanced education and their extensive links to Industry Canada, the granting councils and other federal bodies may, for a period, become secondary to a federal–provincial process that excludes them.

Post-Secondary Education and Canadian Federalism

How has Canadian federalism shaped post-secondary education policy? This simple question is remarkably difficult to answer. As Fred Fletcher and Don Wallace have noted: 'To analyse the impact of federalism – the pattern of divided jurisdiction – on policy outcomes is a difficult task. It involves a speculative activity in which the analyst compares what did (or did not) happen with what might have happened under a system of central dominance or one of provincial dominance' (Fletcher and Wallace 1985, 134). In this 'speculative activity,' federalism is only one of several forces at work. Its independent influence is hard to determine.

Another problem is the diversity of post-secondary issues, stakeholders, and policies. Federalism's influence varies across issues, thus making generalizations difficult. Moreover, federalism's impact, to a degree, is also a function of observers' ideas about the characteristics of a 'good' post-secondary education system. For example, an advocate of a post-secondary education system with an explicit hierarchy of institutions, topped by a select group of research universities, might frown on federalism. Similarly, someone who wants broad access and low tuitions might come to a different judgment. Finally, three of the world's oldest federations, Australia, Canada, and the United States, have high-quality post-secondary education systems that engage both levels of government. Federalism certainly has consequences. It has not immobilized its governments or its post-secondary institutions.

As compared to a hypothetical unitary Canada, federalism, with its active federal and provincial governments, has not led to the development of a coherent national system of student admissions, credit transfers, and student financial assistance. Provincial governments, backed with substantial administrative capacity and constitutional authority, are dominant in these matters. Federalism's diversity trumps national standards. Canada has no national mechanism for planning enrolments, for assessing programs, or for planning infrastructure needs.

Student financial assistance is an important area that is visibly shaped by federalism. The status quo, replete with interprovincial policy differences, has few defenders and many critics. Canada has a complex system that does not meet the needs of a growing student body in an era of economic change, increasing demand for post-secondary education, and rising costs. Many reform proposals have been advanced. Recently, former Ontario premier Bob Rae studied Ontario's post-secondary education system. His report raised serious concerns about the effectiveness of student financial assistance in Ontario. Rae's complex solution called for an income-contingent repayment scheme and major grants as well as loans (Cameron 2005b, 285–6). He wanted Ottawa to provide assistance for living expenses while Ontario would support tuition and fee-related costs. This vision, if deemed superior to the status quo and politically acceptable, at the very least demands federal agreement and an explicit decision by Ontario to leave the Canada Student Loans Program. Other provinces, most recently Quebec, which operates its own plan, have discussed income-contingent systems but no national consensus has emerged. Provinces differ considerably in their views about funding priorities, the policy purposes to be pursued, and the balance between loans and grants, to cite just a few issues. They also differ about how tuitions should be set and students' personal contribution to their education. Canadian federalism limits radical change in an area that, in a hypothetical unitary Canada, could easily be transformed by a determined government.

University research policy shows that federalism, contrary to its stereotype as a rigid, veto-laden system, can move decisively under certain circumstances. Beginning in 1997, the Government of Canada made sweeping policy changes that strengthened Canadian universities' research capacity. It moved without prior discussion with provincial governments and without extensive consideration of the full impact of change on post-secondary education. These changes, at a time when provincial governments were pursuing restraint, quickly made research a higher university priority than it would have been otherwise.

Second, as compared with a hypothetical unitary Canada, research and teaching roles are weakly integrated within institutions. Federal policymakers envision universities primarily as research institutions. Provincial governments see them differently to a considerable degree. Their focus is on more bread-and-butter issues like programs, tuitions, and system quality.

Under these circumstances, federalism institutionalizes a struggle over priorities but allows no easy solution. Another way to look at federalism's impact is to consider the idea that Canada's universities should more clearly differentiate themselves. Under Canadian federalism in the millennium, it is at least possible to envision a system of 'national' universities dedicated to excellent research and graduate studies. An underpinning system of equally excellent undergraduate teaching universities and colleges, on the other hand, would be much harder to achieve without major policy changes.

The Future

The future of post-secondary education policy has recently spawned many thoughtful analyses (Bok 2005; Rhodes 2001; Shapiro 2005; Slaughter and Rhoades 2004). Well-known issues include problems raised by reproductive technologies, the need for broader access, and growing costs. Other concerns, a sample from a very long list, are the impact of communications technologies, the need to 'internationalize' education, and more responsive governance.

Keith Banting (2006) puts these matters in a different light. His point is that the 'welfare state' is much diminished in coverage and ambition. Governments now see education and training as the primary basis of citizen security in the face of economic change. Regrettably, rhetoric has outstripped effective policy. Income support has been cut before major reforms to education have been undertaken. Ingrained inequalities remain that will not be easily reduced by 'better and more education' unless complemented by substantial income assistance. 'A social policy premised on human capital is likely to fail, even on its own terms, if growing economic inequality is followed by growing educational inequality. We are led inevitably back to a debate about redistribution' (Banting 2006, 445).

Banting's analysis has serious implications for post-secondary education. Universities, in particular, may have to change in major ways. For one thing, they will have to think more carefully about their relationships with elementary and secondary education. They will have to do this as an institutional priority, not as an intellectual matter for their faculties of education. 'Lifelong learning,' a staple of university rhetoric, will have to become a genuine priority. To be fair, universities have not been designed to serve society as alternatives to a diminished welfare state. Their ambition is to participate fully in the 'new research paradigm.' They envision themselves as contributors, through research, to the *transformation* of the economy, not as instruments of a reformulated welfare state. Universities have considerable autonomy in their operations and can resist policy initiatives at variance with their priorities.

Can Canadian federalism adapt to post-secondary education's heavy burden? A determined federal government can certainly have a decisive impact. But aggressive federal policy has been unilateral – without provincial participation, advance consent, or direct input, and partial in scope – that is, dealing primarily with research, with universities, and with student financial assistance. On the other hand, the provinces, outside of Atlantic Canada, have done little to foster interprovincial collaboration or to engage Ottawa beyond bursts of lobbying. Governments share the policy space, and operate actively within it, but ultimately do so without reference to shared goals or priorities. An uneven system has resulted. A strong national research system coexists with assorted programs of student financial assistance, varied system 'architecture' in each province, and heterogeneous tuition policies.

An expanded federal role remains the dream of many 'stakeholders' in post-secondary education who want predictability, uniformity, and national standards. A dominant federal presence in education is unlikely. Most obviously, no federal government has manifested imperial ambitions for education. Moreover, the provinces are not likely to give ground. Alberta, flush, until the 2008–09 recession, with large budget surpluses, has great aspirations for its post-secondary education system, Ontario is cantankerous, and Quebec provincial politics are in flux.

Well-known interprovincial differences of interest, wealth, and partisanship preclude cooperation. The Social Union Framework Agreement has a promising vision of intergovernmental collaboration that, whatever its conceptual value, has not yielded dividends.

Might Canadians themselves demand post-secondary education reform given their experiences and aspirations in a changing economy? 'Reform from below' is not likely. Canadians are concerned about access to high-quality education. But public engagement has not gone beyond intermittent concerns about rising tuition and sporadic complaints about the inaccessibility of major universities. The issues are complex, debates are expert-dominated, Canadians are (too) deferential to post-secondary leaders, and, unlike health care, lack of access to post-secondary education is not immediately life-threatening. In the millennium's first decade, strong labour demand in western Canada continues to provide many unskilled or low-skilled persons with jobs at relatively high pay. The short-term good life of a booming economy deflects citizens from thinking about their longer-term interests and needs. Under these circumstances, Canadian governments and educators will remain in the driver's seat, where their options will be structured by federalism. Whether they can deliver is an open question.

Conclusions

This chapter examined the federal–provincial dimension of Canadian post-secondary education policy and funding. As we have seen, this larger realm of policy and public spending can impact on research and innovation policy and university–federal government relations in many direct and indirect ways as priorities and institutions change regarding the overall component parts of both post-secondary education and federalism. In crucial ways higher education's various policy elements are major competitors for large portions of provincial and federal government budgets.

The analysis has stressed the 'uncoordinated entanglement' of provincial and federal policy. Ottawa dominates research funding and plays a substantial role in student financial assistance. The provincial governments, for their part, have constitutional responsibility for post-secondary education and wield extensive powers. Each province provides operating and capital financing for post-secondary education and major programs of student financial assistance. The provincial governments also set tuition fees and have large supervisory and regulatory roles.

Yet, while deeply interdependent, federal and provincial government policy in post-secondary education is not well coordinated. Post-secondary education policy is also characterized by convulsive periods of decisive federal action and considerable interprovincial policy differences.

The chapter also examined the potential impact of a Conservative federal government on post-secondary education policy and federalism. It notes that universities were apprehensive about the 2006 election of a Conservative minority government. Concerns remain but in 2007–08 the Conservative government sent generally reassuring signals to universities and asserted a major federal role in post-secondary education.

Third, the chapter considered the impact of federalism on post-secondary education policy by contrasting the status quo with a hypothetical unitary Canada. Two conclusions emerged. First, federalism has impeded a substantial overhaul of student financial assistance. Second, federalism has contributed to an imbalance in university priorities between teaching and research. Since 1997, aggressive federal government programs for university research have tilted university priorities toward research and made research more important than it would be in either a unitary Canada or a Canada where the provincial governments dominated post-secondary education policy.

Finally, the chapter probed the future of federalism and post-secondary education and reached two conclusions. First, the federal and provincial governments must better coordinate their post-secondary education policies. Demands for substantially increased access to higher education, greater citizen mobility, and the complexity of universities themselves suggest that changes are needed. Second, post-secondary education policy, despite its importance, remains dominated by the interests of governments and post-secondary institutions themselves. This fact makes major changes unlikely. In the decade ahead, post-secondary education issues will continue to push federalism to its limits.

REFERENCES

AUCC. (2007). *Federal Budget Summary, March 19, 2007.* Association of Universities and Colleges of Canada.

Aucoin, Peter. (2003). 'Independent Foundations, Public Money and Public Accountability: Whither Ministerial Responsibility as Democratic Governance.' *Canadian Public Administration* 46 (1): 1–26.

Bakvis, Herman. (2007). 'The Knowledge Economy and Post-Secondary Education: Federalism in Search of a Metaphor.' In *Canadian Federalism: Performance, Effectiveness and Legitimacy*, 2nd ed., ed. Herman Bakvis and Grace Skogstad, 205–22. Don Mills, Ont.: Oxford University Press.

Banting, Keith G. (2006). 'Dis-embedding Liberalism? The New Social Policy Paradigm in Canada.' In *Dimensions of Inequality in Canada*, ed. David A. Green and Jonathan R. Kesselman, 417–52. Vancouver: University of British Columbia Press.

Bok, Derek. (2005). *Our Underachieving Colleges: A Candid Look at How Much Students Learn and Why They Should Be Learning More*. Princeton, N.J.: Princeton University Press.

Cameron, David M. (2003). 'The Challenge of Change: Canadian Universities in the Twenty-First Century.' *Canadian Public Administration* 45 (2): 145–74.

– (2004). 'Collaborative Federalism and Post-Secondary Education: Be Careful What You Wish For.' Paper prepared for the John Deutsch Institute for the Study of Economic Policy, Queen's University, Kingston, Ontario.

– (2005a). 'Post-Secondary Education and Research: Whither Canadian Federalism?' In *Taking Public Universities Seriously*, ed. F. Iacobucci and Carolyn Tuohy, 277–92. Toronto: University of Toronto Press.

– (2005b). 'Ontario's Rae Report: Investing in Growth.' *Canadian Public Administration* 48 (2): 280–7.

Canada. (1951). Royal Commission on National Development in the Arts, Letters and Sciences. *Report*. Ottawa: King's Printer.

– (1958). Royal Commission on Canada's Economic Prospects. *Final Report*. Ottawa: Queen's Printer.

– (1985). Royal Commission on the Economic Union and Development Prospects for Canada. *Report*. Ottawa: Queen's Printer.

– (2002). *Framework of Agreed Principles on Federally Funded University Research: An Agreement between the Government of Canada and the Association of Universities and Colleges of Canada*. Ottawa: Industry Canada.

– (2005). *Fiscal and Economic Update*. Ottawa: Department of Finance Canada.

– (2006a). *Budget 2006: Focusing on Priorities*. Ottawa: Department of Finance Canada.

– (2006b). *Restoring Fiscal Balance in Canada: Focusing on Priorities*. Ottawa: Department of Finance Canada.

Clark, Ian D. (2003). 'Comments on "The Challenge of Change: Canadian Universities in the Twenty-First Century" by David M. Cameron.' *Canadian Public Administration* 45 (3): 401–21.

Council of the Federation. (2006). Advisory Panel on Fiscal Imbalance. *Reconciling the Irreconcilable: Addressing Canada's Fiscal Imbalance*. Ottawa: Council of the Federation.

Council of Ontario Universities. (2006). *Facts and Figures 2006: A compendium of Statistics on Ontario Universities*. Toronto: Council of Ontario Universities.

Fletcher, Frederick J., and Donald C. Wallace. (1985). 'Federal-Provincial Relations and the Making of Public Policy in Canada.' In *Division of Powers and Public Policy*, ed. Richard Simeon, 125–205. Toronto: University of Toronto Press in co-operation with the Royal Commission on the Economic Union and Development Prospects for Canada.

Prichard, J. Robert S. (2000). *Federal Support for Research and Higher Education in Canada: The New Paradigm*. Ottawa: Killam Annual Lecture.

Rhodes, Frank H.T. (2001). *The Role of the American University*. Ithaca and London: Cornell University Press.

Ruch, Richard S. (2001). *Higher Ed, Inc.: The Rise of the For-Profit University*. Baltimore and London: The Johns Hopkins University Press.

Shapiro, Howard T. (2005). *A Larger Sense of Purpose: Higher Education and Society*. Princeton, N.J.: Princeton University Press.

Simeon, Richard, and Ian Robinson. (2004). 'The Dynamics of Canadian Federalism.' In *Canadian Politics*, 4th ed., ed. James Bickerton and Alain C. Gagnon, 101–26. Peterborough, Ont.: Broadview Press.

Slaughter, Sheila, and Gary Rhoades. (2004). *Academic Capitalism and the New Economy*. Baltimore and London: Johns Hopkins University Press.

Smith, Jennifer. (2004). *Federalism*. Vancouver: University of British Columbia Press.

Teichler, Ulrich. (2007). *Higher Education Systems*. Rotterdam: Sense Publishers.

Tupper, Allan. (2003). 'The Chrétien Governments and Higher Education Policy: A Quiet Revolution in Canadian Public Policy.' In *How Ottawa Spends, 2003–2004: Regime Change and Policy Shift*, ed. G. Bruce Doern, 105–17. Don Mills, Ont.: Oxford University Press.

3 Higher Education Funding and Policy Trade-Offs: The AUCC and Federal Research in the Chrétien-Martin Era

CLARA MORGAN

As the two previous chapters have shown, a variety of new federal funding programs and initiatives have been instituted in the last decade to provide a steady flow of research funding to Canadian universities. The purpose of this chapter is to analyse the role that the Association of Universities and Colleges of Canada (AUCC) played in influencing the trajectory of federal research and related policies on universities in the last decade.[1] It also shows some of the tensions and collaborations between the AUCC and other stakeholder interests such as the Canadian Association of University Teachers (CAUT) and the Canadian Federation of Students (CFS), and among the AUCC's university member institutions (for example among research-intensive universities). Overall it also shows, as context for the chapters in part 2 of the book, that the political process for research and innovation plays out in a pool of limited budgetary resources where other aspects of higher education compete with the demands of researchers.

In the case of the AUCC's role, the politics of the research role is necessarily and especially linked to other federal policies in the realm of student assistance. This is because, for the AUCC and the federal government, not to mention individual universities, research policies are often cast as benefits for elites, whereas student assistance is more

---

1 The author is grateful for the assistance provided by Robert Best, Vice-President, National Affairs, with the Association of Universities and Colleges of Canada; David Robinson, Associate Executive Director, with the Canadian Association of University Teachers; Duncan Watt, Vice-President, Finance and Administration, with Carleton University; Dr Robert Rosehart, past President of Lakehead University and Wilfrid Laurier University; and Dr Peter Adams, former Member of Parliament for Peterborough. All errors and omissions are solely those of the author.

readily linked to the politics of the average voter. Thus federal research policies per se can only be suitably explained through some understanding of where and how the overall higher education funding and policy trade-offs are made as between elite interests and broader voter interests centred on students and parents.

The chapter also shows how the AUCC was not just a lobbyist in Ottawa but also became, as chapter 1 has already suggested, a part of the federal–university research governance apparatus through the framework agreements it signed with the federal government on behalf of its university membership.

The chapter explores the changes that were made to federal higher education policies and the AUCC's lobbying efforts in the last decade (see Table 3.1). Once it was able to secure a steady government funding source for research and for higher education, the AUCC sought to preserve the status quo of the significant gains made in the last decade. In contrast, other interest groups such as the Canadian Federation of Students (CFS) and the Canadian Association of University Teachers (CAUT) have challenged some of the federal government's research and related policies on universities and have continued to advocate for increases in federal transfers to the provinces. These alternative views will also be referred to on a more selective basis.

The chapter discusses how the AUCC became an influential voice in higher education and how it crafted a strategy that ensured the continuity and involvement of the federal government in higher education. When the drastic cuts came in the early 1990s, even though the AUCC and other interest groups were shocked by this turn of events, the AUCC was able to mobilize and take a leadership position. The AUCC projected a consensus view on federal university funding but, as the chapter reveals, there were areas of significant disagreement among university presidents. The AUCC membership was diverse and the needs of the more research-intensive larger universities, particularly those of the 'Group of 13,' at times overshadowed those of the smaller much less research intensive universities. The chapter describes how key policy actors helped translate the AUCC agenda into federal policy initiatives. Among these actors was the Liberal Government's Caucus on Post-Secondary Education and Research (GCPSER), which played an important role in helping to communicate and reinforce the AUCC strategy. The chapter concludes with an overview of the effects of federal policies on universities.

Table 3.1
Key AUCC Research and Related Federal Policy Advocacy Realms

AUCC Advocacy	Government Program/Initiative	Budget	Amount
Student Assistance	Canada Millennium Scholarship Fund	1998 – Canada Opportunities Strategy	$2.5 billion endowment. The foundation provides $300 million annually in bursaries and scholarships to 100,000 students.
	Canada Study Grants	1998 – Canada Opportunities Strategy	Provided though Canada Student Loans Program. Canada Study Grants of up to $3,000 a year are made available to full- or part-time students in financial need who have children or other dependants.
Research Funding			
(a) Research Infrastructure	CFI	1997	$800 million up front; $180 million annually. Additional funding provided in subsequent budgets.
(b) Human Resources in Research	Increased funding to Granting Councils	1997	Funding for research grants, scholarships, and fellowships for graduate and post-graduate students.
	Canada Research Chairs	2000	$900 million over five years through the granting councils to establish and sustain 2,000 Canada Research Chairs by 2004–05.
	Canada Graduate Scholarships	2003	An annual cost of $105 million when fully phased in. The new program will complement Canada Research Chairs program.
(c) Direct Costs of Research	Increased research funding	1998–2005	Increased funding to Granting Councils.
(d) Indirect Costs of Research	One-time funding for indirect costs	2001	A one-time investment of $200 million through the granting councils to Canada's universities.
	Permanent indirect costs program established	2003	$225 million per year through the granting councils beginning in 2003–04. Amount increased in subsequent budgets.

The AUCC: Representing the Interests of Universities

Today, the AUCC is a national organization that represents the diverse interests of ninety-four Canadian public and private not-for-profit universities and university-degree level colleges. The AUCC was founded in 1911 when twenty representatives from fifteen Canadian universities met at McGill University to discuss the upcoming Imperial Universities Congress to be held in London in 1912 (Pilkington 1980). Originally known as the National Conference of Canadian Universities (NCCU),[2] the NCCU's Constitution in 1944 described its mission as 'the promotion of higher education in all its forms in Canada' (quoted in Pilkington 1980, 12). After its inception, the NCCU received little 'public visibility' until 1939 when it became involved in helping the federal government with its war effort (Pilkington 1980, ix). Pilkington notes in her history of the AUCC that the 'model of cooperation and collaboration' that was pursued by the federal government and the universities 'has not been achieved since' World War II (1980, 25).

Throughout the 1940s and '50s, the NCCU advocated increased federal university funding. It receded from the public eye as the federal government relinquished its higher education commitments to the provinces whilst at the same time retaining its funding research commitment. The only area that the AUCC could successfully lobby for at the federal level was research funding. The AUCC maintained a 'federally oriented position' (Pilkington 1980, 232) since, as past AUCC President Morgan described it, there was a 'federal dimension for universities' (ibid., 291) which could not be addressed at the provincial level. The AUCC's strategy in the late 1970s and '80s was to build a closer working relationship with the federal government and its departments (Pilkington 1980).

The AUCC's ascendance as a significant and influential voice in national higher education is a relatively recent phenomenon. The turning point for the AUCC took place in the late 1980s, when it responded to

---

2 The federal government committed in 1956 to increased funding to higher education but it wanted the NCCU to administer educational grants so that it would not be seen as intervening in provincial affairs. In order for the NCCU to legally take on this responsibility, it had to be incorporated. It also was facing some internal organizational issues when it created another body called the Canadian Universities Foundation (CUF). The AUCC was incorporated on April 3, 1965 through an Act of Parliament. This resolved the problems it faced with administering grants and reorganized the NCCUC (now with Colleges)-CUF into one entity (Pilkington 1980).

its membership's demand for the AUCC to become an effective voice for higher education. A primary instigator for this demand was the looming threat of the disengagement of the federal government in postsecondary education, particularly since the Established Programs Financing (EPF) arrangements were designed to reduce cash transfers to the provinces. This situation provided the AUCC with an opportunity for putting forth new policy alternatives.

New Strategic Directions for the AUCC

For the AUCC and its membership, a decline in higher education funding or, even worse, the elimination of federal higher education funding would translate into universities having to compete entirely with other provincial institutions for provincial funding. In order to implement a strategy of becoming an effective voice for higher education, the AUCC set in motion a renewal and restructuring process. This entailed building the AUCC's credibility as a reliable source for research and analysis in higher education. The AUCC needed to develop its capacity for in-house advocacy, for research, and for informed policy analysis. In the early 1990s, it set out to hire the right professional staff to achieve these objectives. With this expertise on board, the AUCC could develop a more coherent set of higher education policy recommendations that it could communicate to various departments of the federal government and to politicians.

An important outcome of the AUCC's restructuring process was the establishment of three advisory committees to the Board of Directors: the Standing Advisory Committee on Funding (today known as the Standing Committee on Finance), the Standing Advisory Committee on University Research, and the Standing Advisory Committee on International Relations. The Standing Advisory Committee on Funding, which was chaired by David Smith, Principal of Queen's University, produced a report that summed up the AUCC's strategic orientation and outlined its advocacy agenda. At its core, this advocacy agenda sought to maintain the continuity of federal funding in higher education and to ensure increased federal funding for higher education. The report's key elements remain relevant to AUCC's current strategic orientations.

From the standpoint of the AUCC, its optimal approach was to recommend new policy directions for higher education funding. The AUCC recommendations that were developed in 1992 were based on two underlying principles. The first was the recognition that 'higher education and research are intrinsically linked to the economy and the prosperity of a country and to the social and economic well-being of its citizens' (AUCC 1992, 16). The second principle was the reaffirmation of Quebec as a distinct society and the reflection of this in 'federal-provincial arrangements relating to postsecondary education' (AUCC 1992, 16).[3]

The AUCC's new strategic directions for influencing federal involvement in higher education were initially pursued along two policy tracks – student assistance and research funding. As it was about to pursue its advocacy campaign, the AUCC and other stakeholders involved in higher education were dealt a bombshell: the Liberal government drastically reduced its spending in the mid-1990s, including its expenditure on higher education. In the next section, I will describe federal retrenchment policies and the universities' reactions to these cuts.

Federal Retrenchment

The Liberals' election campaign program of 1993 recognized that deficit reduction had to be addressed, but this agenda item was not initially a priority for the Liberals. Once in power, the Liberals were surprised to find that they had inherited a 'bigger than expected deficit' (Feehan 1995, 33). Ottawa was faced with a large government debt which was continuously growing. The primary concerns for the federal government became getting its spending under control and reducing the deficit (Maslove 1996). Paul Martin, the Finance Minister at the time, was forced to take drastic measures. A retrenchment policy was pursued in the areas of government transfers to the provinces and federal program spending.

In October 1994 Lloyd Axworthy, Minister of Human Resources Development, released his Green Paper on social policy. The Green Paper proposed a phasing out of higher education funding in the next three to four years and 'folding' it into a student income repayment scheme (Cameron 2002, 157). To compensate for the funding shortfall, most universities would have to resort to tuition increases. Because of the negative reception that the Green Paper received, the federal government changed its proposals. Instead of scrapping

---

3 The second principle needs to be seen in light of the constitutional debates that Canada was undergoing in the early 1990s and the failure of the Meech Lake Agreement. At the same time, it is important to recognize that the AUCC is one of the few large and diverse national organizations that has maintained full Quebec membership.

Established Programs Financing (EPF), it extended it to incorporate the Canada Assistance Plan (CAP) and branded it with a new name, the Canada Health and Social Transfer (CHST). What the CHST effectively did was to reduce federal financial transfers by $4.6 billion between 1995–96 and 1997–98 (Haddow 1998, 114). Not only did the federal government cut its transfer payments to the provinces, it also reduced its funding of other programs such as research funding to the granting councils.

These budget cuts translated into reductions in university and college revenues and in research funding to the granting councils. Even though universities had begun to seek other sources of revenue in the 1980s, this became even more of a priority in the mid-1990s (Tudiver 1999). To compensate for the shortfall in funding, many universities and colleges increased their tuition costs.[4] Universities also sought money in the form of donations from the business community (some signing donor agreements), they fundraised, they developed business–university partnerships, and they signed exclusive rights contracts for selling products to university students (see Tudiver 1999; Turk 2000).

The Universities Mobilize

The events of the mid-1990s were characterized as a 'crisis' in higher education as universities were deeply affected by the federal government's policy of retrenchment. Several interest groups mobilized in reaction to this policy and came together to address the crisis. Among these groups was the AUCC, which, as mentioned earlier, was pursuing an advocacy policy that focused on student assistance and research funding. Another policy actor that reacted to the funding cuts was a group of ten research-intensive universities – or the 'Group of 10'[5] – which came together to formulate their own strategic approach to federal involvement in higher education. The Group of 10 felt that the AUCC was not moving quickly enough in addressing the research funding needs of universities.

In 1995, the AUCC appointed a new president to lead it during these turbulent times. Robert Giroux, who had extensive federal government

---

4 Tuition was regulated in Quebec, British Columbia, and Manitoba. Other provinces allowed for tuition increases but imposed ceilings (Tupper 2003).
5 The group has expanded to include thirteen universities and is today more commonly know as the 'Group of 13.'

experience as secretary of the Treasury Board and comptroller general of Canada, was brought in because of his 'thorough understanding of the complex policy environment in which universities are operating' (AUCC 1995a). As Howard Tennant, chair of the AUCC, noted: 'In these times, when universities are concerned about tough issues such as accountability, performance indicators, and new funding formulas, it is perfect to have found an individual who understands not only public policy issues but also the university's roles of teaching, research and scholarship' (ibid.).

Meanwhile, as noted earlier, the major research universities in Canada mobilized and formed their own lobby group. The Group of 10, composed of Canada's largest and/or longest-established universities, such as the University of Toronto, McGill, Queen's University, and UBC, was interested in lobbying the federal government for research dollars and felt that the AUCC was moving too slowly on this advocacy item. The AUCC found itself in a very difficult position, as it was trying to build a common-ground foundation among its diverse membership. The AUCC discouraged the Group of 10 from publicly taking a formal position, as this would weaken the AUCC's ability to lobby on behalf of all of its membership and would reflect a discordant voice among universities. At the same time, the emergence of the Group of 10 pressured the AUCC into focusing its advocacy efforts on university research funding.

With Robert Giroux on board, the AUCC had a savvy leader who could navigate the federal landscape and translate the AUCC's interests into federal policy language. Giroux worked closely with Robert Prichard, the president of the University of Toronto, to develop the AUCC's strategic plan. The next section examines the role that that the AUCC played in the development of the Liberal government's federal higher education policies. The discussion points to how the AUCC distanced itself from advocating for student assistance and eventually ended up focusing its energy on the university research agenda.

Student Assistance

Before the budget cuts were announced in 1994, the AUCC's initial recommendations for student assistance involved two funding mechanisms. The first was Income Contingent Repayment loans (ICR), and the second entailed scholarships and bursaries. The rationale for ICR was to loan students the money for their education costs. Once they had completed their education, these graduates would place a fraction of their earnings towards loan repayment.[6] With the unfolding of the Liberal government's retrenchment policy and the negative reception given to Axworthy's Green Paper, the AUCC adjusted its strategy. It began to move away from recommending ICR loans as a form of student assistance. Its approach was to continue to recommend improved funding for student assistance and to underscore the importance of equitable access to higher education (AUCC 1995b).

The AUCC initiated the Student Assistance Round Table as part of its original strategy. In late 1996, the AUCC convened a meeting that brought together the AUCC, the Association of Canadian Community Colleges, the Canadian Alliance of Student Associations, the Canadian Association of Student Financial Aid Administrators, the Canadian Association of University Teachers (CAUT), the Canadian Federation of Students (CFS), and the Canadian Graduate Council. These representative organizations worked together to develop a 'package of proposals' that were outlined in the Round Table's report, *Renewing Student Assistance in Canada: The Student Assistance Reform Initiative* (AUCC 1997a). Released in January 1997, the report recommended targeted grants to needy students, deferred grants to help students manage their loan repayments, a work-study program so that students would have access to 'non-repayable aid,' and, 'tax measures ... to encourage savings for education and speedier payback loans' (AUCC 1997a).

As the government began to experience a budget surplus in 1997, Prime Minister Chrétien announced the government's intention to establish a Canada Millennium Scholarship Endowment Fund in his 'Address in Reply to the Speech from the Throne.' In response to this development, the Student Assistance Round Table updated its *Renewing Student Assistance in Canada* and issued a new document in November 1997. In it, the members of the Round Table reiterated the same proposals as in the January 1997 report. They also addressed the creation of needs-based grants through the Millennium Scholarship Fund and developed a list of proposals for flexible loan repayment (AUCC 1997b).

The efforts of the Student Assistance Round Table and their strategy for finding common ground paid off when the federal government

---

6 Even though ICR loans, in principle, were not intended to fund increases in tuition costs, they 'almost always' were put forth 'in conjunction with a proposal to increase the proportion of costs of a higher education charged to students through tuition fees' (Cameron 1995, 169).

presented its 1998 Budget and unveiled the Canadian Opportunities Strategy. Many of the Round Table's recommendations found their way into the Canadian Opportunities Strategy in one form or another. These included the creation of the Canada Millennium Scholarship Foundation, a private, independent organization; the creation of Canada Study Grants; improvements to the Canada Student Loans Program; and tax relief for interest on student loans, extended repayment period and interest relief period, and reductions in the loan principal.

The Canadian Millennium Scholarship program was not warmly received. The Bloc Québecois viewed it as interfering with Quebec's jurisdiction over education. According to the Bloc, this funding should be returned as part of the transfer payments that were cut to the provinces. The Canadian Federation of Students (CFS) advocated for the replacement of the Millennium Scholarship program with a national program for needs-based grants. The federal government was criticized for not coordinating this federal scholarship program with provincial student assistance programs. This became an example of what Tupper refers to in chapter 2 as the federal government's 'uncoordinated entanglement.' These unilateral federal actions resulted in some provinces cutting back on their own student assistance funding (CAUT 2004b). For example, the government of Nova Scotia used CMS funding to eliminate its loan remission program (CAUT 2004b). In 2006, the Canada Millennium Scholarship Foundation signed agreements with the provinces which involved funnelling federal money to the provinces. These monies are later disbursed by the provinces.

Since it was created with a ten-year mandate, the Canada Millennium Scholarship Fund came up for renewal under the Conservative Government in 2008. The Conservatives did not renew the fund but opted to increase access to PSE through other programs. The Fund had provided some students with access to post-secondary institutions and helped to contain their debt load. Yet the majority of Canadian students continue to face several barriers such as high tuition costs, high debt loads, and high debt repayment interest rates. As the Foundation's Executive Director and CEO has stated, 'debt remains a major concern, both for the 60 per cent of 18- to 24-year-olds who pursue some post-secondary studies and for any efforts to increase Canada's rate of post-secondary participation in the years to come' (CMSF 2006).

In later advocacy campaigns, the AUCC did not pursue the issue of student assistance. It focused its efforts on obtaining a steady source of research funding for universities from the federal government. The AUCC has not publicly critiqued the current system of student assistance. Since universities depend on student tuition fees to help fund their operating expenditures, they would prefer to have a guaranteed revenue source rather than no revenue at all.

The financial situation of Canada's system of higher education is complex and cannot be addressed by a series of federal funding initiatives. Statistics indicate that Canadian universities and colleges are able to generate 45 per cent of their revenue largely from tuition costs and other sales of goods and services. The other 55 per cent of their revenue comes from government transfers. It is important to note that the provinces provide more than 80 per cent of all government transfers ($15 billion in 2007), whereas the federal government only allocates about 16 per cent ($3 billion in 2007). In 2007, the expenditures of universities and colleges amounted to $34 billion, with a reported deficit of about $200 million (Statistics Canada 2007). As Tupper has discussed in chapter 2, the Council of the Federation's study on the 'fiscal imbalance' has recommended a dedicated post-secondary transfer to the provinces at $4.9 billion, with a 4.5 per cent annual escalator as a solution to addressing the provincial financing of higher education.

At the institutional level, universities are faced with a difficult financial situation and have had to compromise their capacity to deliver a quality education to their students. Approximately 75 per cent of university expenditures are allocated to salaries and benefits. On average, these expenses are rising annually by 5 per cent. Universities are not able to increase their revenues sufficiently in order to finance these costs. At the same time, universities have seen student enrolments rise to 1 million students in 2004–05 (Statistics Canada 2006). They have had to increase teacher/student ratios and resort to hiring sessional instructors in order to minimize their costs. It is these realities that university leaders, administrators, academics, and students have to deal with on a daily basis.

Research Funding

As noted earlier, the crafting of the AUCC's university research agenda came about with the appointment of Robert Giroux as president in 1995. Giroux's experience and seniority in the federal government provided the AUCC with the strategic know-how for moulding the AUCC agenda to the federal and political machinery. Among the university presidents, it was Robert Prichard, President of the University of Toronto, who articulated clearly the need for increased federal funding for university research.

Broadly speaking, AUCC's strategic directions for research funding were encompassed in two briefs submitted to Parliament. The first, submitted in October 1996, was entitled *Putting Knowledge to Work: Sustaining Canada as an Innovative Society: An Action Agenda*, and the second, submitted in September 1997, *Sustaining Canada as an Innovative Society: An Action Agenda*. Several groups collaborated in producing these briefs, including the AUCC, the Canadian Association of University Teachers (CAUT), the National Consortium of Scientific and Educational Societies, the Canadian Consortium for Research, the Humanities and Social Sciences Federation of Canada, and the Canadian Graduate Council. To sustain Canada as an innovative society, the briefs recommended 'promoting research careers; arresting the erosion of Canada's research infrastructure; and enhancing partnerships to foster knowledge and technology flows' (AUCC, CAUT, and the National Consortium of Scientific and Educational Societies 1996). By the year 2000, the AUCC had honed its message and identified four criteria for research funding: 1) research infrastructure; 2) human resources in research; 3) direct research costs; and 4) indirect research costs.

RESEARCH INFRASTRUCTURE

Sustaining Canada as an Innovative Society proposed the creation of an Infrastructure for Innovation Program (IIP), which was to be aimed at 'modernizing the core equipment and physical assets of Canada's university research enterprise' (AUCC et al. 1996). The agenda also highlighted the importance of long-term sustenance of the research infrastructure, which requires the funding of 'overhead costs' associated with research (ibid.). Ottawa addressed the need for investment in research infrastructure by creating the arm's-length Canada Foundation for Innovation (CFI) as a non-profit corporation. Various institutions and individuals were consulted during the implementation of the CFI, including AUCC's President Robert Giroux and other university presidents. An informal advisory group was set up for this process.

As the Lopreite and Murphy analysis in chapter 5 shows, CFI's mandate was to strengthen the capacity of Canadian universities, colleges, research hospitals, and non-profit research institutions to carry out world-class research and technology development that benefits Canadians (CFI 2005). CFI could only fund up to 40 per cent of a project's infrastructure costs, with the remaining 60 per cent funded by other partners from the private, public, and voluntary sectors.

The AUCC has been a strong advocate of the CFI. It has strongly supported it as 'an effective model to invest in state-of-the-art university research infrastructure' (AUCC 2005a). What is especially important for the universities is CFI's ability to plan over a multi-year time frame. This accommodates the needs of the research community, which requires 'long-term planning horizons and funding assurances' (AUCC 2005a). Without such a mechanism in place, the government's yearly appropriations in a normal budget cycle would not be able to meet the long-term planning needs of the research community.

In contrast to the AUCC, the CAUT has been concerned with the impact that the CFI funding has had on the commercialization of universities and on exacerbating regional and institutional disparities. The CAUT (2004b) pointed to the fact that a business partner that provides 60 per cent financing towards an infrastructural investment can exercise a veto over the choice of the project. This situation forces universities to seek projects that are aligned with private needs rather than aimed at the public interest. The CAUT has also pointed to the inequity caused by this co-funding formula, which favours richer provinces and urban centres over poor provinces and rural areas. The richer provinces and urban centres receive a larger share of funding simply because they have access to a greater number of partners[7] (CAUT 2004b).

HUMAN RESOURCES IN RESEARCH

Sustaining Canada as an Innovative Society pointed to the brain drain and attributed it to the 'shortage of challenging career opportunities' in Canada (AUCC et al. 1996). In order to stem this outflow of university graduates, the agenda argued for the creation of 'stimulating research environments' by proposing a New Research Frontiers Program to support the research work of young faculty members and a Transition Awards program to provide opportunities for university graduates to work outside of academia (AUCC et al. 1996). The federal government responded to the demand for promoting research careers by expanding its graduate student support programs, establishing the Canada Research Chairs program and creating the Canada Graduate Scholarship program.

In 1999, Ottawa set aside $900 million to establish two thousand Canada Research Chairs (CRC) in universities across Canada. The

---

7 Since the Atlantic provinces were at an economic disadvantage in being able to raise money as compared with other Canadian provinces, Ottawa created the Atlantic Innovation Fund in 2001 'to strengthen the economy of Atlantic Canada by accelerating the development of knowledge-based industry.' The fund 'focuses on R&D projects in the area of natural and applied sciences, as well as in social sciences and humanities, where these are explicitly linked to the development of technology-based products, processes or services, or their commercialization' (ACOA 2005).

program objectives included strengthening Canada's research capacity by attracting and retaining the best researchers, improving the training of personnel through research; improving universities' capacity to generate and apply new knowledge, and promoting the best possible use of research resources (CRC 2006a). Universities who submit nominations to this program can also request infrastructural support from CFI (CRC 2006b).

The AUCC views the Canada Research Chairs (CRC) program as a successful investment which has contributed to Canada's 'brain gain' (AUCC 2006a). Other interest groups, such as the CAUT, have raised concerns about the program. The CAUT has conducted its own review of the program and found several issues of concern (CAUT 2005). These have included problems of inequity in the allocation and distribution of research chairs. The CAUT has lobbied for both the continuation and the reform of the CRC program, since universities have grown to depend on the CRC as a source of revenue for faculty members' salaries. In 2006, the Canadian Human Rights Commission settled a complaint launched by eight female university professors in 2003 against the CRC's discriminatory structure. The settlement ensures that the CRC program is structured equitably and is non-discriminatory (CAUT 2006).

The federal government also encouraged the growth of human resources in research by expanding its scholarship programs through the creation of the Vanier Canada Graduate Scholarships. Funding for both these programs is available through the three granting councils. Other programs, such as the Undergraduate Student Research Awards, are aimed at stimulating interest in research and in encouraging undergraduates to pursue graduate studies and research careers in engineering and the natural sciences (NSERC 2005).

DIRECT COSTS OF RESEARCH

As chapters 1 and 2 have shown, direct research costs are funded through the grants received by university researchers from the research councils. The Liberal government's retrenchment policy had also reduced the research councils' budgets. One of the points that the AUCC consistently made to Ottawa was that the research councils' budgets were too low when compared with those of other national governments, in part because they did not include research overheads (AUCC 1999). By framing research funding within the context of Canada's international competitiveness, the AUCC was able to drive home the importance of federal involvement in this area and the need for increased funding. Since 1998, as Doern's analysis in chapter 4 shows, Ottawa has steadily increased its funding to the federal research granting councils.

The AUCC has readily supported increases in direct costs funding without critically examining the repercussions of these policies. Polster (2007) has looked at how the growing dependency of universities on research grants as a source of income is transforming university relations. These transformations are significant, since they alter the relations between and among various actors in higher education in complex ways.[8] She has also assessed the negative implications of the current research grants regime, including the undermining of the 'university's mission, namely teaching, administrative service, and public service' (Polster 2007, 614). Pocklington and Tupper (2002) have also reflected on the university's mission and have criticized the prioritization of research over teaching.

INDIRECT COSTS OF ACADEMIC RESEARCH

A key aspect in research funding that remained to be addressed was the funding of indirect costs.[9] Initially, the AUCC lobbied for infrastructure funding that included overhead or operating costs. With the creation of the Canadian Foundation for Innovation and the funding of new university research, universities found that their operating costs also increased. The AUCC needed another funding source for these overhead costs, which it labelled as 'indirect costs of research'. In effect, as Tupper in chapter 2 notes, indirect costs were similar to operating grants to universities. To successfully campaign for indirect costs funding, the AUCC produced several analytical reports on how indirect costs were

---

8 For example, relations among academics are changed as certain professors are rewarded for their ability to obtain research grants; university administrators take a more active role in encouraging academics to apply for research grants and in recruiting team members who can successfully get grants; and the federal government indirectly exercises more control over university research directions as university administrators align their priorities for research with those of the granting councils.
9 The AUCC (2006b) defines indirect costs of research as 'general costs associated with operating and maintaining facilities and resources (e.g., laboratories, libraries and computer networks), managing the research process (e.g., research coordination, grant applications and management of intellectual property) and regulation and safety compliance (e.g., human ethics issues, animal care, biohazards and environmental assessment).'

funded in other jurisdictions, such as the United States and the United Kingdom. In its briefings and presentations to various Senate committees, the AUCC pointed to Canada's 'comparative disadvantage' when it came to funding research costs because it did not have a mechanism for funding indirect costs (AUCC 1999).

Ottawa announced the creation of a one-time indirect costs payment by allotting $200 million to seventy-nine degree-granting institutions in its 2001 budget. From the AUCC's perspective, this was an indication that Ottawa recognized that indirect costs were a significant component of research funding. AUCC's past president, Robert Giroux, described the federal government's decision as a 'winning strategy' and as a 'much-needed and far-sighted initiative' (AUCC 2001). The indirect costs program was extended in 2002 and became a permanent program. It received $325 million in the 2009–10 budget (Indirect Costs Program 2009).

The CFS and the CAUT responded with considerable ambivalence to the creation of the indirect costs program. For the CFS, this source of funding was seen as inadequate because it does not replace the deep cuts that universities incurred as a result of the reduction in transfer payments. From the CFS perspective, students remain vulnerable to increased tuition costs since universities continue to be under-funded (CFS 2006). The CAUT has also cautioned that indirect costs funding should not be viewed as replacing core university funding (CAUT 2002).

Federal research money was directed at new research funding but not at addressing university maintenance costs. As the Canadian Association of University Business Officers (CAUBO) demonstrated in their 2000 report *A Point of No Return: The Urgent Need for Infrastructure Renewal at Canadian Universities*, universities required $3.6 billion in deferred maintenance funding, of which $1 billion was urgently needed (CAUBO 2000). This funding would go towards fixing dangerous structures and repairing or upgrading ventilation systems, underground tunnels, and leaky roofs (Chester 2000). As these were not attractive or flashy investments, universities faced difficulties raising money for such costs from the private sector (ibid.). Because the AUCC was regarded as having expertise in governmental relations, CAUBO went to the AUCC for assistance in lobbying the government. CAUBO was able to communicate its report findings by appearing in front of the Standing Senate Committee on National Finance in 2001.

The AUCC embraced the federal funding programs in higher education and university research and did not publicly oppose the push towards public–private research partnerships. It appears that the AUCC had not thought through the repercussions that these policies might have on the university research environment. For example, the CFS has pointed out that with university researchers increasingly becoming involved in private sector partnerships, there have been reports of research misconduct. Students who observe such misconduct have been afraid to come forward since they are provided with little protection (CFS 2006a). The academic freedom of university professors and researchers has to be protected. As demonstrated by the famous case of Nancy Olivieri, researchers who take an ethical position towards industry-sponsored research come under attack and risk losing their jobs. In fact, not only did the Hospital for Sick Children not provide Olivieri with any support as she struggled against the pharmaceutical company Apotex, it also 'was negotiating with Apotex for a huge financial donation' (Schafer 2007, 112). Incidentally, since the AUCC does not directly represent hospitals unless they are owned and operated by universities, this can pose problems for the AUCC in terms of their public statements on these particular kinds of controversies.

Despite these troubling effects of industry-sponsored university research, the AUCC, as we shall see in the next section, welcomed the government's innovation agenda and signed an agreement that further encouraged universities to pursue the path of the commercialization of university research.

The AUCC and Canada's Innovation Strategy

The federal government launched its innovation strategy in February 2002 by releasing two documents: *Achieving Excellence: Investing in People, Knowledge and Opportunity* and *Knowledge Matters: Skills and Learning for Canadians*. In *Achieving Excellence*, Ottawa called on Canadians 'to work together to improve innovation performance' (Industry Canada 2002). In its document *Knowledge Matters*, Ottawa set key milestones and goals to be achieved in various areas, including higher education. Of particular concern to the federal government were accessibility to higher education for all Canadians and excellence in higher education (HRDC 2002).

In response to the government's call for involvement in the implementation of its innovation strategy, the AUCC issued in July 2002 an Action Plan entitled *A Strong Foundation for Innovation*. In this Action Plan, the AUCC underlined the major contributions universities make in the areas of innovation and knowledge. The Action Plan identified the 'collective commitments that universities are prepared to make as their contribution to the innovation agenda.' For the AUCC, these collective commitments also required 'complementary actions by all major stakeholders,' including government (AUCC 2002, 2).

Not all the higher education interest groups were willing to embrace the government's innovation agenda. For example, the CAUT was worried that the strategy narrowly defined research and innovation in terms of new products for the market. Such an approach, the CAUT cautioned, could marginalize research that was not commercially lucrative. The CAUT's response has been and continues to be to underline the need for core university funding (CAUT 2002).

The Framework of Agreed Principles on Federally Funded University Research

From the AUCC's perspective, the federal government needed to commit to certain key areas of investment. These included indirect costs funding at 'an overall rate of 40 per cent of direct research funding' (AUCC 2002, 5) and increased funding in direct costs, infrastructure, student assistance, and scholarship programs. In turn, the federal government needed assurance from the universities that they would follow through on their commitments. In order to formalize the working relationship between these stakeholders, the Government of Canada and the AUCC developed what came to be known as the 'Framework of Agreed Principles on Federally Funded University Research.'

The Framework was signed on 18 November 2002 and announced at the National Summit on Innovation and Learning. Proponents of the agreement viewed it as a testimony to Ottawa's commitment to higher education funding and its recognition that Canada's economic success and competitiveness depended on the knowledge and research generated by universities and colleges. The universities, on their part, committed to doubling the amount of research they generate and tripling their commercialization performance. The CAUT's president warned that such a commitment could end up privileging commercial research over basic research.

In the Framework, the AUCC also committed to delivering a performance report on university research and knowledge transfer. The AUCC published its first report, entitled *Momentum: The 2005 Report on University Research and Knowledge Transfer,* in October 2005. It simultaneously launched a website, Momentum, which not only provided access to the report but also highlighted the variety of research areas that universities are undertaking.

By signing the Framework of Agreed Principles with the federal government, the AUCC has asserted itself as an intrinsic component of the higher education and research governance apparatus. The trade-off for universities has been a commitment to pursuing the commercialization of research. In the next section, we shall see how the strengthening of the AUCC as a lobby group in this political arena could not have been achieved without the mediation of the Government Caucus on Post-secondary Education and Research.

Inside the Governing Liberal Party: A Lobby Group for Universities and Research

A major challenge the AUCC faced as an advocate for increased funding and federal involvement in higher education was the fragmented nature of the academic community. There were many disparate voices communicating with Ottawa in the early 1990s. In fact, the federal government tended to dismiss higher education in favour of other sectors because of this lack of coherence, particularly in the face of more concerted lobbying in numerous other policy fields. In reality, members of the academic community shared many concerns and were advocating along similar lines but they were unable to speak with one voice. Because of this weakness, legitimate and significant issues relating to higher education were not being communicated to Ottawa. Liberal Party members who were also members of the academic community recognized that there was a need for representation of the higher education community inside the then governing Liberal party, particularly since the budget cuts of the mid-1990s had adversely affected the higher education and research communities (McKarney 2003).

In 1994, Peter Adams, former Member of Parliament for Peterborough and professor emeritus at Trent University, and John English, former Member of Parliament for Kitchener and professor at the University of Waterloo, created a caucus committee known as the Government Caucus on Post-secondary Education and Research (GCPSER). When it was first created, the Caucus repeatedly informed the academic community that they needed to speak coherently to the federal government. In order for the Caucus to successfully communicate higher education priorities inside government, the academic and research communities needed to find areas of mutual concern.

The GCPSER was an invaluable forum, as it provided members of the academic and research communities with an unprecedented opportunity to begin the process of finding common ground. For example, Robert Giroux, president of the AUCC, and Don Savage, then Executive Director of the Canadian Association of University Teachers (CAUT), appeared together in front of the GCPSER. This initial encounter was an important milestone because it paved the way for future encounters with other groups around issues of student assistance and research funding.

The GCPSER, made up of twenty MPs and two Senators, was under the jurisdiction of the Liberal Party's Social Caucus and conducted annual public hearings at National Caucus meetings. Throughout the year, the GCPSER met with representatives of the academic and research communities (Adams 2001). The AUCC was one of the groups that the Caucus regularly consulted with. Based on these consultations, the GCPSER provided a consensus view of higher education priorities to the Minister of Finance.

Adams, who chaired the GCPSER, described the Caucus as a 'lobby group within the system' which 'advocated for postsecondary and research interests' (McKarney 2003). The GCPSER has also lobbied hard for research funding in the areas that the AUCC has advocated, including the Canadian Foundation for Innovation (CFI), the Canadian Millennium Scholarship Foundation, and the Canada Research Chairs (McKarney 2003). Of concern to the GCPSER was ensuring the permanency of federal funding in higher education. For the GCPSER, publicly endowed foundations such as the CFI and CMS Foundation were policy instruments that secured long-term funding of PSE and research. The GCPSER was crucial in assisting the AUCC in advancing its agenda and in communicating the various policy alternatives to the right people within the governing Liberal party, including the Prime Minister and the Cabinet.

Conclusions

This chapter has examined the role that the AUCC has played in putting forth various federal research and related higher education policy alternatives during the Chrétien-Martin era. It has shown some of the tensions and collaborations between the AUCC and other stakeholder interests such as the Canadian Association of University Teachers and Canadian Federation of Students and among the AUCC's university member institutions. And it has described the interesting role of the

GCPSER as a 'lobby from within' the governing Liberal Party caucus during the Chrétien-Martin Liberal era. The chapter has demonstrated that the AUCC was not just a lobbyist in Ottawa, but also became a part of the federal–university research governance apparatus through its framework agreements with the federal government on behalf of its member universities.

The focus of the chapter has ultimately been on federal research policies, but in the case of the AUCC's role, the politics of the federal research role is necessarily linked to other federal policies on higher education, particularly in the realm of student assistance. This is because, for the AUCC and the federal government, not to mention individual universities, research policies are often cast as benefits for elites, whereas student assistance is more readily linked to the politics of the average voter, especially students and their parents. Thus federal research policies per se and the AUCC's role as lobbyist can only be fully explained through some understanding of where the overall higher education funding and policy trade-offs are made, as between elite interests and broader voter interests centred on students and parents.

The analysis has shown that as an interest group representing the quite diverse and often conflicting interests of universities, the AUCC has worked hard at securing a constant flow of federal money for research to its membership. Today, federal university funding goes to four areas of research which the AUCC describes as the 'four pillars of university research': ideas, talent, infrastructure, and institutional support (AUCC 2006c). These translate into direct costs of research, human resources, research infrastructure, and indirect costs of research.

This chapter reinforces Tupper's conclusions in chapter 2 that what is problematic about the present state of federal involvement in federal research and higher education is that these federal policies blur the lines of responsibility in higher education between the federal and provincial levels of governance. The federal government has traditionally been responsible for funding research whereas provincial governments were responsible for the operating costs of universities. As Wolfe notes, recent federal research funding has created an environment for more 'targeted and applied research funding at the expense of the broader-based support for basic research provided by the granting councils,' and these kinds of support give the federal government 'significant leverage over the allocation of both research and teaching priorities within universities' (Wolfe 2005, 332). Such a targeted approach to research funding is associated with the belief that this funding will pay off and that it will translate into economic opportunities for commercial growth and for industrial innovation.

The AUCC has taken a non-confrontational and sometimes uncritical approach in its advocacy of federal research funding. Its press releases, presentations, and briefs indicate that it was happy with the Liberals' higher education policies. The AUCC has incorporated the federal government's national economic goals as part of its own arguments for research funding (Tupper 2003). Other interests groups such as the CAUT and CFS have been quite vocal in highlighting the shortcomings of federal research policies for universities. The CAUT's Executive Director has characterized the CRC, the CFI, and the Millennium Scholarship Fund as 'putting a fancy porch on a crumbling building,' since these programs do not address the chronic underfunding universities continue to experience (Grant 2002, 262).

Whether one is a proponent or a critic of federal research and related policies with respect to universities, what has to be kept in mind is that these policies have altered the way universities and academics do their work. Direct federal interventions in higher education have had significant effects on the day-to-day workings of academia. Some of these effects include increased involvement of the private sector in university research, increased competition among universities and academics for research grants, the emergence of commercialization as a university priority, and the gradual distancing of the university from its core mission of serving the public interest.

From the AUCC's vantage point, what has been achieved is a solid foundation – the groundwork for a healthy university and research environment. Yet the AUCC needs to reflect upon and assess the foundation that has been built, as future policies will only exacerbate some of the negative effects of the policy outcomes that are being detected today. James Downey, keynote speaker at the AUCC's 2003 annual meeting, warned his colleagues about the consequences of 'the warm embrace of economic functionalism' (Downey 2003, 7) and reminded them of the university's mission:

> The primary mission of the university is not to train but to educate, not to do research or transfer technology, not to prepare students for jobs but to make them more discerning people, capable of seeing through the political and commercial hucksterisms of their times, of establishing their own values and finding their own meaning in life, of constructing and expressing their own compelling narratives. Through teaching and research the university must

cultivate a spirit of intellectual dissent. Not for its own sake, but in the interests of a free, tolerant, enlightened, and improving society. (Ibid., 16)

REFERENCES

ACOA. (2005). 'Atlantic Innovation Fund, Program Overview.' Atlantic Canada Opportunities Agency. Available from: http://www.acoa.ca/e/financial/aif/over.shtml (cited 30 November 2006).

Adams, P. (2001). *Annual Report, 2000–2001: Into the New Century.* Available from: http://www.peteradams.org/publications/annualreports/annualreport20002001.htm (cited 30 November 2006).

AUCC. (1992). *Federal Support for University Education and Research.* Ottawa: Association of Universities and Colleges of Canada.

– (1995a). 'A New President for the Association of Universities.' 15 September. Ottawa: AUCC. Available from: http://www.aucc.ca/publications/media/1995/09_15_e.html (cited 13 November 2007).

– (1995b). 'Notes for a Presentation to the House of Commons Standing Committee on Finance regarding Bill C-76, The Budget Implementation Act.' 9 May. Ottawa: AUCC. Available from: http://www.aucc.ca/publications/reports/1995/budgetact_05_09_e.html (cited 30 November 2006).

– (1996). 'Backgrounder: Federal Budget 1996: Impact on Universities.' 7 March. Ottawa: AUCC. Available from: http://www.aucc.ca/publications/media/1996/03_07_e.html (cited 30 November 2006).

– (1997a). 'Renewing Student Assistance in Canada: The Student Assistance Reform Initiative.' 20 January. Available from: http://www.aucc.ca/publications/reports/1997/stuaid_renew_01_20_e.html (cited 30 November 2006).

– (1997b). 'Renewing Student Assistance in Canada II: Key Elements.' 17 November. Available from: http://www.aucc.ca/publications/reports/1997/stuaid_key_11_20_e.html (cited 30 November 2006).

– (1998). 'Notes for a Statement to The House of Commons Standing Committee on Finance regarding Bill C-36: An Act to implement certain provisions of the budget tabled in Parliament on 24 February 1998.' Available from: http://www.aucc.ca/publications/reports/1998/billc36_04_30_e.html (cited 30 November 2006).

– (1999). Research and Education: The Underpinnings of Innovation: Canada as a Knowledge-based and Innovative Society.' September 17. Ottawa: AUCC. Available from: http://www.aucc.ca/publications/reports/1999/prebudbrief_09_17_e.html (cited 30 November 2006).

- (2001). 'Payment for Indirect Costs a Winning Strategy.' December 10. Ottawa: AUCC. Available from: http://www.aucc.ca/publications/media/2001/12_10_e.html (cited 30 November 2006).
- (2002). *A Strong Foundation for Innovation: An AUCC Action Plan*. July. Ottawa: AUCC. Available from: http://www.aucc.ca/_pdf/english/reports/2002/innovation/innoactionpl_e.pdf (cited 30 November 2006).
- (2005a). 'Universities Reaffirm Support for Canada Foundation for Innovation.' 16 February. Ottawa: AUCC. Available from: http://www.aucc.ca/publications/media/2005/02_16_cfi_e.html (cited 30 November 2006).
- (2005b). 'Universities Applaud Increased Support to Postsecondary Education and University Research in Economic Update.' November 14. Ottawa: AUCC. Available from: http://www.aucc.ca/publications/media/2005/11_14_e.html (cited 30 November 2006).
- (2005c). *Momentum: The 2005 Report on University Research and Knowledge Transfer*. Ottawa: AUCC. Available from: http://www.aucc.ca/momentum/en/report/ (cited 30 November 2006).
- (2006a). 'Notes for a Presentation to the House of Commons Standing Committee On Finance.' 27 September. Available from: http://www.aucc.ca/_pdf/english/reports/2006/prebudget_speaking_notes_09_27_e.pdf (cited 30 November 2006).
- (2006b). 'Indirect Costs of Federally Funded University Research.' Available from: http://www.aucc.ca/_pdf/english/reports/2006/indirect_costs_10_19_e.pdf (cited 30 November 2006).
- (2006c). The Four Pillars of University Research. Available from: http://www.aucc.ca/_pdf/english/reports/2006/4_pillars_research_10_19_e.pdf (cited 30 November 2006).

AUCC, CAUT, and the National Consortium of Scientific and Educational Societies. (1996). 'Putting Knowledge to Work: Sustaining Canada as an Innovative Society. An Action Agenda.' 18 October. Ottawa: AUCC. Available from: http://www.aucc.ca/publications/reports/1996/knowwork_10_18_e.html (cited 3 October 2006).

Buchbinder, H., and J. Newson. (1990). 'Corporate-University Linkages in Canada: Transforming a Public Institution.' *Higher Education* 20 (4): 355–79.

Cameron, D. (1995). 'Shifting the Burden: Liberal Policy for Post-Secondary Education.' In *How Ottawa Spends, 1995–96: Mid-Life Crisis*, ed. S. Phillips, 121–36. Ottawa: Carleton University Press.

- (2002). 'The Challenge of Change: Canadian Universities in the Twenty-First Century.' *Canadian Public Administration* 45 (2): 145–74.
- (2004). 'Post-Secondary Education and Research: Whither Canadian Federalism.' Paper presented at the University of Toronto conference,

Taking Public Universities Seriously, 3–4 December. Available from: http://www.utoronto.ca/president/04conference/downloads/cameron.pdf (cited 30 November 2006).

CAUBO. (2000). 'Backgrounder – "A Point of No Return: The Urgent Need for Infrastructure Renewal at Canadian Universities."' Canadian Association of University Business Officers. April. Available from: http://www.aucc.ca/_pdf/english/media/caubo_e.pdf (cited 13 November 2007).

CAUT. (2002). 'Canada's Innovation Agenda: CAUT's Response.' Available from: http://www.caut.ca/en/issues/commercialization/innovation agendaresponse.pdf (cited 30 November 2006).

– (2004a). 'Statement to the House of Commons Standing Committee on Finance regarding the 2005 Federal Budget.' Available from: http://www.caut.ca/en/publications/briefs/2004financebrief.pdf (cited 30 November 2006).

– (2004b). 'CMS Foundation Seeks Program Feedback.' May 2004. Available from: http://www.caut.ca/en/bulletin/issues/2004_may/other%20news/article_1_cms.asp (cited 15 November 2007).

– (2005). *Alternative Fifth Year Review of Canada Research Chairs Program*. Ottawa: CAUT. Available from: http://www.caut.ca/en/publications/briefs/2005_crc_review.pdf (cited 30 November 2006).

– (2006). 'Settlement Welcomed in Federal Research Chairs Discrimination Complaint.' 9 November. Available from: http://www.caut.ca/en/news/comms/20061109crchrcomplaint.asp (cited 30 November 2006).

CFI. (2005). 'CFI Overview.' Available from: http://www.innovation.ca/en/about-the-cfi/cfi-overview (cited 30 November 2006).

CFS. (2005). 'Canadian Federation of Students' Submission to the House of Commons Standing Committee on Finance October 2005.' Available from: http://www.cfs-fcee.ca/html/english/research/submissions/sub-2005-financectteebrief.pdf (cited 30 November 2006).

– (2006a). 'Canadian Federation of Students' Submission to the House of Commons Legislative Committee on Bill C-2.' 29 May. Available from: http://www.cfs-fcee.ca/html/english/research/submissions/C2_submission-EN.pdf (cited 15 November 2006).

– (2006b). 'Campaigns and Lobbying: Research and Graduate Funding.' http://www.cfs-fcee.ca/html/english/campaigns/researchandgrad.php (cited 30 November 2006).

– (2007). *Strategy for Change: Money Does Matter*. Canadian Federation of Students. October. Available from: http://www.cfsadmin.org/quickftp/Strategy_for_Change_2007.pdf (cited 14 November 2007).

Chester, B. (2000). 'Canadian Campuses in Creaky Condition.' *McGill Reporter*, 27 July. Available from: http://www.mcgill.ca/reporter/32/18/infrastructure (cited 13 November 2007).

CMSF. (2004.) 'Seeks Program Feedback.' Canada Millennium Scholarship Foundation. May. Available from: http://cfsadmin.org/quickftp/Strategy_for_Change_2007.pdf (cited 14 November 2007).

– (2006). *Fast Forward. Annual Report 2006*. Canada Millennium Scholarship Foundation. Available from: http://www.millenniumscholarships.ca/uploadfiles/documents/annual_reports/2006_ar_en.pdf (cited 15 November 2006).

CRC. (2006a). 'About us.' Canada Research Chairs. Available from: http://www.chairs.gc.ca/web/about/index_e.asp (cited 30 November 2006).

– (2006b). 'Program Details: Nominate a Chair.' Canada Research Chairs. http://www.chairs.gc.ca/web/program/nominate_e.asp (cited 30 November 2006).

CSLP. (2005). 'About the Canada Student Loans Program (CSLP).' http://www.hrsdc.gc.ca/asp/gateway.asp?hr=en/hip/cslp/About/01_ab_MissionProgram.shtml&hs=cxp (cited 26 October 2005).

Department of Finance. (1998). *Budget 1998: The Canadian Opportunities Strategy* Available from: http://www.fin.gc.ca/budget98/cos/cose.pdf (cited 30 November 2006).

– (2005). *The Budget Plan 2005*. Available from: http://www.fin.gc.ca/budget05/pdf/bp2005e.pdf (cited 30 November 2006).

Downey, J. (2003). 'The Consenting University and Dissenting Academy: Binary Friction.' Speech to AUCC Annual Meeting, 6 April 2003.

Feehan, J. (1995). 'The Federal Debt.' In *How Ottawa Spends, 1995–96: Mid-Life Crises*, ed. S. Phillips, 31–58. Ottawa: Carleton University Press.

Grant, K. (2002). 'A Conversation on the Future of the Academy with James Turk, PhD, Executive Director, Canadian Association of University Teachers.' *The Canadian Review of Sociology and Anthropology* 39 (3): 261–74.

Haddow, R. (1998). 'How Ottawa Shrivels: Ottawa's Declining Role in Active Labour Market Policy.' In *How Ottawa Spends, 1998–99: Balancing Act: The Post-Deficit Mandate*, ed. Les Pal, 99–126. Ottawa: Oxford University Press.

HRDC. (2002). *Knowledge Matters: Skills and Learning for Canadians*. Human Resources and Development Canada. Available from: http://www11.sdc.gc.ca/sl-ca/doc/summary.pdf (cited 30 November 2006).

Indirect Costs Program. (2009). 'About the Program.' Available from: http://www.indirectcosts.gc.ca/about/index_e.asp (cited 30 July).

Industry Canada. (2002). *Achieving Excellence: Investing in People, Knowledge and Opportunity*. Ottawa: Industry Canada. Available from: http://www.innovationstrategy.gc.ca/gol/innovation/site.nsf/vDownload/Page_PDF/$file/achieving.pdf (cited 30 November 2006).

Kinder, J. (2003). 'The Doubling of Government Science and Canada's Innovation Strategy.' In *How Ottawa Spends, 2003–04. Regime Change and Policy Shift*, ed. G. Bruce Doern, 204–20. Don Mills, Ont.: Oxford University Press.

Kingdon, J. (1984). *Agendas, Alternatives, and Public Policies*. Boston: Little Brown.

Maslove, Allan. (1996). 'The Canada Health and Social Transfer: Forcing Issues.' In *How Ottawa Spends, 1996–97: Life Under the Knife*, ed. Gene Swimmer, 283–301. Ottawa: Carleton University Press.

McKarney, L. (2003). 'Behind the Scenes of the Budget.' *Science Magazine*, 4 April. Available from: http://sciencecareers.sciencemag.org/career_development/issue/articles/2310/behind_the_scenes_of_the_budget (cited 30 November 2006).

NCE. (2005). 'About Us.' Networks of Centres of Excellence. Available from: http://www.nce.gc.ca/about_e.htm (cited 30 November 2005).

Newson, J. (1998). 'The Corporate-Linked University: From Social Project to Market Force.' *Canadian Journal of Communication* 23 (1): 107–15.

NSERC. (1999). *Report on Plans and Priorities: 1999–2000 Estimates*. Ottawa: Treasury Board Secretariat.

– (2005). Undergraduate Student Research Awards. Available from: http://www.nserc.gc.ca/sf_e.asp?nav=sfnav&lbi=1a (cited 30 November 2006).

Pilkington, G. (1980). 'Speaking with One Voice: Universities in Dialogue with Government.' Unpublished paper. McGill University.

Pocklington, T., and Allan Tupper. (2002). *No Place to Learn: Why Universities Aren't Working*. Vancouver: University of British Columbia Press.

Polster, C. (2003/04). 'Canadian University Research Policy at the Turn of the Century: Continuity and Change in the Social Relations of Academic Research.' *Studies in Political Economy* 71/72: 177–99.

– (2007). 'The Nature and Implications of the Growing Importance of Research Grants to Canadian Universities and Academics.' *Higher Education* 53 (5) (May): 599–622.

Schafer, A. (2007). 'Commentary: Science Scandal or Ethics Scandal? Olivieri Redux,' *Bioethics* 21 (2): 111–15.

Statistics Canada. (2006). 'Universities and Colleges Revenue and Expenditures.' Available from: http://www40.statcan.ca/l01/cst01/govt31a.htm (cited 15 November 2007).

– (2007). 'University Enrolments by Registration Status and Sex, by Province.' Available from: http://www40.statcan.ca/l01/cst01/educ53a.htm (cited 15 November 2007).

Tudiver, N. (1999). *Universities for Sale: Resisting Corporate Control over Canadian Higher Education*. Toronto: James Lorimer.

Tupper, Allan. (2003). 'The Chrétien Government and Higher Education: A Quiet Revolution in Canada Public Policy.' In *How Ottawa Spends, 2003–04:*

Regime Change and Policy Shift, ed. G. Bruce Doern, 105–17. Don Mills, Ont.: Oxford University Press.

Turk, J. (2000). *The Corporate Campus. Commercialization and the Dangers to Canada's Colleges and Universities*. Toronto: James Lorimer.

Wolfe, D. (2005). 'Innovation and Research Funding: The Role of Government Support.' In F. *Taking Public Universities Seriously*, ed. Iacobucci and C. Tuohy, 316–40. Toronto: University of Toronto Press.

PART TWO

Research and Innovation Policy Issues, Impacts, and Relations

4 The Granting Councils and the Research Granting Process: Core Values in Federal Government–University Interactions

G. BRUCE DOERN

The purpose of this chapter is to examine core values in the interactions between the federal government and universities in the research granting process through a comparative analysis of the evolution of the Natural Sciences and Engineering Research Council (NSERC), the Canadian Institutes of Health Research (CIHR), formerly the Medical Research Council, and the Social Sciences and Humanities Research Council (SSHRC), and through an examination of recent changes in federal policy concerns and capacities.[1] The focus is very much on the granting process rather than on research infrastructure via the Canada Foundation for Innovation (examined in chapter 5) or the joint conduct of research between universities and government (examined in chapter 8). While the focus in this chapter is on relationships between the federal government's granting councils and universities in an overall sense, there are also further particular relationships between universities and academic researchers that exist and that are being changed. These, too, will be referred to where appropriate.

The four core values examined are the ones set out in the second part of chapter 1's analytical framework, namely: 1) the relative independence of researchers from government; 2) the peer-reviewed nature of university research; 3) the federal need for accountability and value for money in the use of public funds; and 4) the Federal Government and

---

1 I wish to thank Jeff Kinder, Joan Murphy, Les Pal, Bob Slater, Allan Tupper, Eliot Phillipson, Jac van Beek, Jerome Doutriaux, and James Meadowcroft for their constructive and helpful comments on earlier drafts of this chapter or in general discussion on the issues covered. Several federal government officials also offered very useful comments on previous drafts and so I am indebted to them as well both for this assistance and for discussions at early research stages.

Canadian society's need for objective and useful social, economic, and policy-relevant research in a fast-changing Canadian and global context. There are, of course, many other focal points for assessing the nature of research grants and the granting process but a focus on these core values is often ignored, as are the overall relationships between the federal government and the granting bodies in the granting process.

The analysis is based on a review of relevant published literature and reports (academic and governmental), including reports by the three granting councils. These primary sources were complemented by some discussions with staff at the three granting councils and in selected federal departments and agencies.

The structure of the chapter is reasonably straightforward. The first section briefly profiles the core mandates of the three granting councils and their recent evolution as organizations. This is followed by a closer look at the four core values in the federal government–university research granting relationship, founded not only on granting councils' needs and pressures but also on the federal government's own needs for policy research and accountability. Conclusions then follow.

The Three Granting Agencies: Core Mandates and Basic Changes

This section describes and analyses the origins and recent evolution of the three main federal research granting councils, the Natural Sciences and Engineering Research Council (NSERC), the Canadian Institutes of Health Research (CIHR), and the Social Sciences and Humanities Research Council (SSHRC).[2] It sets out in a broad way each granting council's origins and mandate and its evolving and current status as a granting body and federally funded agency. This brief historical overview is also centred on how each council relates to the broader evolution of federal science and technology (S&T), innovation, education, and health policies (de la Mothe 2003; Kinder 2003; Canada 1996, 1999a; Beesley et al. 1998).

The Natural Sciences and Engineering Research Council of Canada (NSERC)

NSERC was established in 1978, when it grew out of the granting activities of the National Research Council of Canada (Lopreite 2006; NSERC

---

2 Special thanks are due to Débora Lopreite, Russell LaPointe, and Joan Murphy, whose research on the three granting bodies I draw on and cite below, augmented by my own research for this chapter.

1999). The statutory mandate of NSERC is 'to promote and assist research in the natural sciences and engineering other than the health sciences, and advise the Minister in respect of such matters relating to such research as the Minister may refer to the Council for its consideration.'[3] The main objective of the Council's programs is to promote and support both research and the formation of highly qualified personnel for the Canadian system of science and technology. Its main sub-objectives are: a) to support a diversified base of high-quality research in the natural sciences and engineering; b) to assist in the development of highly qualified personnel; c) to promote and support targeted research in selected fields of national importance; and d) to encourage closer links between the university community and other sectors of the economy.

As expressed in 1990, NSERC had three main responsibilities:

- To secure a healthy research base and, as a corollary, to secure a sound balance between the research base and the more targeted programs of research;
- To secure an adequate supply of highly qualified personnel who have been well educated in basic science and trained with 'state-of-the-art facilities';
- To play an active role in facilitating collaboration between R&D performing sectors in Canada, and in providing mechanisms that will improve the transfer of knowledge and technology.[4]

By the late 1980s, NSERC was seeking a greater research focus on developments related to the knowledge-based economy (KBE). As Lopreite notes, 'NSERC emphasized the need for more integration between Canadian universities and the private sector' (Lopreite 2006, 109). NSERC's priorities then and into the early 1990s related partly to three new core technologies that had the potential to transform all industrial sectors: biotechnology, microelectronics, and new materials. These technologies were also gaining increased emphasis in the federal government's competitiveness strategies of the early 1990s (Doern 1996), when the discourse of competitiveness was used rather than the later focus on innovation in the mid- and late-1990s (Lopreite 2006, 109). As Lopreite argues, 'the development of these new technologies affected a mix of research across the R&D spectrum but also across all

---

3 *Natural Science and Engineering Research Council Act, 1976–77*, c. 24.
4 Quoted in Lopreite 2006, 108.

the traditional disciplines of the natural sciences and engineering. In many ways, they were helping to break down some of the boundaries that traditionally separated or distinguished core scientific disciplines' (Lopreite 2006, 109). In this context, as Lopreite further observes, 'the Canadian challenge was to strengthen and maintain its competitiveness in its older economy 'resource' industries (for example, forest and agri-food products), to develop the new technology-intensive industries (for example, telecommunications), and, in the process, to establish also a more ecologically benevolent industrial base under the emerging paradigm of sustainable development' (Lopreite 2006, 109).

By 2002, the innovation agenda was front and centre in overall federal S&T and innovation policy. Lopreite sums up this period, as follows:

> NSERC conducted its own consultation strategy in the spring of 2002 with more than 300 stakeholders, including students, university and college professors and administrators, industry leaders and federal and provincial public agencies. Following these and other consultations, NSERC announced its five year *Investing in People – An Action Plan*. These plans built on past programs but also were clearly geared to dealing with more specific challenges that emerged from the consultations (and from the larger federal innovation agenda). Smaller initiatives were promised to deal with: 1) increasing the pipeline of young people interested in science and engineering, through science promotion activities and increasing the number of undergraduate student research awards; 2) ensuring that Canada develops a skilled and talented labour force to satisfy the anticipated demand for HQP (in part by increasing the value of NSERC postgraduate scholarships but also by studies to determine why research fellows often do not return to Canada); 3) ensuring that Canadian research is world-class and internationally competitive (through measures such as improving its work in the intellectual property management program); and 4) minimizing the time a researcher (or student) spends in applying for funding and peer review funding proposals. (Lopreite 2006, 109)

Following the arrival of the Martin Government, several 'pilot projects' were developed (NSERC 2004, 2004a). These included: 1) research capacity in Canada's smaller universities (which often was a synonym for regional universities); 2) helping community colleges in innovation; and 3) improving science and math education (throughout the education system). The creation of two new NSERC regional offices in Moncton and Winnipeg was also announced. In addition consultation was underway on a framework for dealing with and managing 'Big Science,' to be carried out through the Office of the National Science Advisor (Lopreite 2006, 122; Office of the National Science Advisor 2005). 'Big science' refers to expensive large capital-intensive kinds of science facilities that historically have included telescopes used in astronomy, space research, nuclear research, and the like.

These 2005 plans were premised by David Emerson, the then new Industry Minister, on the view that NSERC (and other agencies) had to 'ensure that the research and development efforts of the universities and government find their way into the market place' (quoted in Lopreite 2006, 122). In these Budget plans NSERC had transformed, at least for the purposes of communication, its older 'people, discovery, and innovation' programs into 'tomorrow's innovators' (people), 'brain gain' (discovery), and 'realizing the benefits' (innovation) (Lopreite 2006, 122).

The above Budget document noted several other challenges, one of which was referred to as the 'culture gap.' The report thus stressed that 'one of NSERC's goals is to broaden acceptance by universities and faculty of the importance and legitimacy of the commercialization of university research results in Canada, given the low level of industrial R&D for economic, cultural and historic reasons' (quoted in Lopreite 2006, 122). Other gaps were also identified, such as the research transfer gap, the receptor capacity gap (in industry), and the skills gap (in areas such as business management). The report again noted changing university demographics, but in this case it was the extraordinary growth in the number of first-time applicants. Thus, past policies were now creating new demands. It also noted that CFI funding which created new equipment and facilities at universities (see chapter 5) now needed to have the HQP and grants to make facilities actually operational (Treasury Board Secretariat 2005).

By 2005–06, NSERC's budget stood at $865 million, with 44.7 per cent going to discovery, 31.4 per cent to people, 19.1 per cent to innovation, and 4.7 per cent to administration. In the period after 1995, NSERC's budget stood at just under $500 million, and then initially declined in the 1995–97 period (the effects of Program Review budget cuts overall) before rising fairly steadily from 1998–99 on as overall federal S&T investment increased. In constant 1995 dollars, the growth figures are reduced. For example, its current dollar budget in 2004–05 was about $800 million, whereas its constant 1995 dollar budget, taking inflation into account, was about $700 million (NSERC 2006).

The Canadian Institutes of Health Research (CIHR)

The CIHR was formed in 2000 by merging the former Medical Research Council of Canada (MRC) and the National Health and Development Research Program (NHDRP) from Health Canada, and by creating in the process thirteen 'virtual reality' research institutes devoted to interdisciplinary research on applied health and wellness problems and challenges (Murphy 2007).

The immediate genesis in the formation of the CIHR was the work of a Task Force comprised of leaders in the Canadian health research community, which met in 1998 and which discussed ways to better link researchers from all disciplines and also ways in which resources could be focused on Canada's major health challenges. The guiding principles for the CIHR are to:

- Adopt research priorities that are linked with Canadian health priorities and complement the provincial investment in research, education, and health;
- Encompass and support the full spectrum of health research – from basic science to clinical research to population health – recognizing the important role of investigator-initiated research;
- Ensure that Canadian researchers succeed in the worldwide research community through the application of peer review as fundamental to the evaluation of research excellence and internationally competitive levels of funding;
- Encourage individual Institutes within the network to conduct unique programs – from capacity-building to third-party partnerships – in pursuit of the improved health and well-being of Canadians;
- Collaborate with all organizations that have demonstrated a capacity to support or conduct health research. CIHR supports and recognizes the major contributions to health research by voluntary health organizations, universities, provincial granting bodies, and individual research centres;
- Recognize and support the central role that universities and associated health science centres play in education, in training, and in creating interdisciplinary research opportunities (Canada 1999b, 9).

In structural terms, the CIHR is, like the former Medical Research Council (MRC), an agency of the federal government reporting to

Parliament through the Minister of Health. The CIHR incorporated the operations of the MRC itself, but its core new structural feature is its series of Institutes. Its initial set of thirteen Institutes emerged in 2000 from further consultation and includes Institutes devoted to traditional or familiar areas such as cancer research, but also several that meet the test of broader interdisciplinary realms, such as the Institute of Healthy Aging and the Institute of Aboriginal People's Health. The thirteen include Institutes for:

- Aboriginal People's Health
- Aging
- Cancer Research
- Circulatory and Respiratory Health
- Gender and Health
- Genetics
- Health Services and Policy Research
- Human Development, Child and Youth Health
- Infection and Immunology
- Musculoskeletal Health and Arthritis
- Neurosciences, Mental Health, and Addiction
- Nutrition, Metabolism, and Diabetes
- Population and Public Health (Canada 2005, 4).

Federal policy stressed in 2000 that 'the Institutes would not be centralized "bricks and mortar" facilities. Instead these virtual organizations would support and link researchers who may be located in universities, hospitals and other research centres across Canada' (Canada 2005, 4). Another key feature of the Institutes is that they 'would support researchers who approach health challenges from different disciplinary perspectives' (Canada 2005, 4). The vision of transformation which the CIHR was to bring saw the existing system and model as having *'dispersed* research efforts, disciplinary *separation, separate* from delivery, and *multiple agendas*,' whereas the new model anchored in the CIHR would be *'integrated*: across geography/institutions; across scientific disciplines; into the health system; and with the national health agenda' (Canada 2005, 6). Joan Murphy's analysis of the CIHR Institutes comments on their role as follows:

The Institutes concept provides researchers with an opportunity to align their research strengths to priority areas in the Institutes. Institutes improve the support for well-developed research communities and may be able to meet many of their goals through the structure of grants for basic and applied research under the CIHR's peer review system. In this way, the Institutes assist in shaping the research environment, while promoting investigator-driven research. Through the organization of the Institutes structure, Institutes with well-established, and well-organized, research communities can also more easily identify and promote key strategic initiatives. The flexibility to support strategic initiatives allows the Institutes to address the priority needs of their researchers and stakeholder communities. Preferably, strategic initiatives should be supported by multi-disciplinary research teams. Others may support efforts to build a research base in emerging fields.

Other Institutes, simply given their focus area, have broad responsibility to encourage the development of accomplished and experienced investigators in research fields that are underdeveloped and have a small research base.

Each Institute's multi-year strategic plan details a proposed mixture of strategic grants. Once approved by the Governing Council, the Institutes are expected to receive a direct allocation to support its strategic initiatives. Institutes can also receive funding for Institutes' development initiatives.

The Institute's budget will also indirectly accrue funding from the investigator-initiated researchers that are assigned to it. All Institutes will be able to count on a floor level for research funding, to support a base amount of research. Moreover, Institutes will have the budgetary flexibility to invest more money in research that is relevant to their approved strategic plan – above and beyond their allocations from the CIHR pool. (Murphy 2007, 19–20)

By its seventh year of operation, the CIHR stated that, through its Institutes, it 'uses a problem-based multi-disciplinary and collaborative approach to health research' (Canada, 2005, 4). Each of these Institutes is also expected to conduct research across the *four pillars of health sciences inquiry (bio-medical, clinical, health systems and services, and population health)*. But it also distinguishes its core research granting business from its work through the Institutes. Thus the larger part of the CIHR budget provides grants based on investigator-driven ideas and proposals through open competitions judged through peer review processes. In a sense this competitions-focused larger portion represents somewhat the same kinds of discipline-based research funded by the former MRC.

With respect to funding, in 1999–2000 the CIHR (MRC) had total funding of $289 million, with $251 million going to open competitions, $24 million to strategic initiatives, and $14 million to operating expenditures. In 2004–05 the CIHR had total funding of $666 million, with $441 million going to open competitions, $178 million to strategic initiatives, and $46 million to operating expenditures (CIHR 2006). Thus, the total budget had more than doubled, the open competitions component had almost doubled but as a percentage of the total budget had decreased from 87 per cent to 66 per cent, and the strategic initiatives component had increased sevenfold and had moved from 8 per cent of the budget to 27 per cent. Operational costs had increased fourfold, largely because managing the virtual-reality Institutes is more complex in transactional terms. It must be stressed that there is no clear-cut way overall to differentiate the proportions of grant spending that go to the Institutes per se and to investigator-driven research per se. This is because there are significant overlaps and hence practical problems of assigning them to one category or the other.

Among the three granting councils examined in this chapter, the CIHR has garnered by far the largest percentage increases in spending in the last few years. Its projected budget for 2007–08 was $788.6 million. In part this was due to the greater inherent political saliency of health issues among voters compared to science and engineering and the social sciences. Its Blueprint 2007 set out five areas of focus:

- Developing National Research Platforms and Initiatives;
- Supporting Strategic Research through CIHR's Institutes;
- Strengthening Canada's Health Research Communities;
- Knowledge Translation; and
- Commercialization (Canada 2005, 2).

The commercialization focus reflects the fact that CIHR's formative years coincided with the emergence of the federal innovation strategy in 2002 and the greater expectation of research that would support a health 'industry' and not just a health sector. The commercialization goals have been a difficult challenge given that the large majority of the health research community sees itself as being in a public interest–centred research endeavour overall rather than a commercial one. The CIHR's work does, however, support through university–industry partnerships significant areas of research regarding biotechnology and research on clinical trials (Murphy 2007).

Some of this larger health focus has kept the CIHR in a good position vis-à-vis the Harper Conservative government. In the 2 May 2006 federal budget it received a 2.4 per cent increase in its budget and in addition was given $21.5 million over five years to fund research into pandemic preparedness (CIHR 2006, 1–2).

The Social Sciences and Humanities Research Council (SSHRC)

The SSHRC was established in 1978 (Miller 2006; LaPointe 2006). Previously, funding for the social sciences and the humanities had been a part of the mandate of the Canada Council, whose focus had been on funding the arts, and the budget of which included both taxpayer funds and endowment funds. The SSHRC is entirely taxpayer funded and its statutory purpose is to 'promote and assist research and scholarship in the social sciences and the humanities' (quoted in LaPointe 2006, 128). As is the case with NSERC, SSHRC reports to Parliament through the Minister of Industry but it is not formally a part of Industry Canada as a department (SSHRC 2004, 3).

SSHRC's main goals are to:

- support such independent research as in the judgment of scholars will best advance knowledge;
- assist in and advise on maintaining and developing the national capacity for research;
- encourage research on themes considered by the Council to be of national importance; and
- facilitate the communication and exchange of research results (LaPointe 2006, 129).

Five core principles are derived from these goals in the SSHRC mandate (SSHRC 2004, 10). These are:

- that it will only fund research that is of excellent quality and meets international standards that are determined by a peer review process which is balanced by factors such as: region, linguistics, gender, discipline, and the size of the university;
- the selection for grants will be a competitive process that is free from governmental and bureaucratic influence;
- funding will not be limited to particular fields, but rather will be open to all fields of research and will be committed to sharing all knowledge and ideas;

- SSHRC is committed to consistently renewing Canada's research capacity;
- SSHRC is committed to the accurate reporting and stewardship of public funds.

In line with these principles, SSHRC has five main program funding areas: 1) basic research; 2) research training; 3) strategic or thematic research; 4) research communication support; and 5) institutional support.

Russell LaPointe's analysis of SSHRC notes that 'throughout the existence of SSHRC there has been criticism by parts of the academic community about thematic research. But other parts also lobby for key themes or get government officials to plant the seed as to why a given topic or issue deserves attention. The criticism has been ... that the Council's decision to move to thematic research meant that the Council was bending to the desires of the federal government' (LaPointe 2006, 133). LaPointe shows that SSHRC funded initial thematic areas such as the Human Context of Science and Technology and Women in the Labour Force, and more recently areas such the New Economy and also CURA grants that focused on community–university research linkages (LaPointe 2006, 133).

In 2004–05, SSHRC began an elaborate process to change itself from a granting council to a 'knowledge council.' It stated:

> What we are aiming for is a new council – one that remains in charge of delivering grants awarded through peer review, but one that also directly supports and facilitates the sharing, synthesis and impact of research knowledge. In short, we are aiming for a knowledge agency. We need to work out concretely what it means to the human sciences to contribute to a knowledge society. Everyone has to take stock, both those who produce knowledge and those who rely on it to do their work effectively. (SSHRC 2004, 10)

LaPointe points to four reasons and/or pressures that were propelling the SSHRC initiative, and to related aspects of the size and demographics of the research community.

> First, there are the changes within the research community which include: demographic changes; an increased movement towards multidisciplinary research; and how research is being done in general. Second, the new focus on research by the federal government that started in the late 1990s has been pivotal. A third factor is a new era of accountability within government expressed through the demonstration of value from public money

and results-based management. And finally, there is the rising demand for research in the social sciences and the humanities.

The research community within the social sciences and humanities is a large one. The SSHRC has consistently had the largest number of people (full-time faculty and graduate students) to support and a broader range of disciplines that fall under their mandate than the other two granting councils. Nonetheless, it has been receiving lower amounts of funding than the other granting councils from the government. The SSHRC's clients come from 90 universities and are over half of the full-time faculty and graduate students. The Council points out that there are a number of demographic changes taking place within theses two groups. Full-time faculty is getting older and being replaced by younger faculty members. The SSHRC points to the increase of 62 per cent in new scholar applications for SRC grants between 1999 and 2003 compared to only an 11 per cent increase in regular scholars over this time. The Council estimates that over the next ten years one third of full-time professors will retire. This means that new researchers will have to be trained to replace them. There is a concern that new researchers are not being hired in the tenure track positions or being funded adequately. With a new generation of researchers the Council recognizes there will be demands and challenges that will be different than their predecessors and they have to be met. (LaPointe 2006, 136)

SSHRC's budget in 2005–06 was $292 million, 45 per cent of which was spent on research, 48 per cent on people, and 7 per cent on knowledge mobilization. Its total grant funding in 1995–96 was just over $91 million. As was the case with NSERC, its budget declined in the 1996 and 1997 periods but then began to rise from 1998–99 on (SSHRC 2006).

The Core Values: A Closer Look

With the above portraits of the three granting bodies as background, this section now takes a second closer look at the four core values previewed earlier and discussed in chapter 1.

The Relative Independence of Academic Researchers from Government

As noted in chapter 1, the implicit contract between the state and the researcher awarded a grant is that researchers have high degrees of independence regarding the choosing of the area or topic of investigation, the theoretical perspective, the methodology used, and the conclusions or findings reported. Indeed, some would argue that underpinning the value of independence is an even deeper value of freedom of inquiry or academic freedom, but for the purposes of this chapter we subsume this in a broad notion of relative independence. Independence is buttressed by the publication of research subject to peer review processes and an overall view that publicly funded research is in the public domain. From the point of view of government, the rest of the contract with researchers is that they will carry out this research professionally, competently, and ethically and also that they will be broadly accountable for the research, practice basic value for money norms and rules, and disseminate their research in the public domain.

The primacy of the relative independence for researchers is a powerful and important norm, but the independence of any major institution is not without limits. Nation states strive for independence, as do the press and mass media, the courts, the church, universities overall, and markets. Science has been cast as the 'fifth estate' in a similar way. But each of these core institutions also know that on an everyday basis they are interdependent and must be accountable to others.

All of the three granting councils are stout defenders of the independence of research from government and, more broadly, of research as a public good. All of their mandate statements stress the independent nature of research, and of course the governing structure of the granting councils reinforces this independence since the board of directors or council members are at arm's length from government and make all decisions about overall research priorities and about how the granting process will work and be structured.

Even where the granting council is administering grants in thematic or policy-relevant areas of research, the councils defend the independence of the actual granting process and of the research subsequently conducted.

But notions of independence from government are also quite nuanced and subject to pressures for greater interdependence and partnering. To assess some of these influences the analysis needs to look further at peer review as a value or norm and also later at the accountability and value for money and the importance placed on the government's need for policy-relevant research.

Peer Review, the Peer Review Process, and Investigator-Driven Research

There is obviously a close relationship between the norm or value of research independence and that of peer review. 'Independence' refers to the features already noted above but these are sharply reinforced by peer review norms and processes, including assessment by other independent experts, the anonymity of assessors/reviewers, and the structure of peer review committees or groupings. The granting body's core relationship, its 'granting' allocative role, is with individual researchers or small teams of researchers who have applied for a grant on a competitive basis with other researchers across Canada. Such researchers suggest and devise the topics, issues, questions, and methodologies, and their applications are assessed through peer review and ranked. Those ranked high by peer reviewers and review committees are given the funds within the limits of the granting agency's overall budget or budget in a particular program. As defined earlier, peer review means review by fellow experts (Canadian and international) in a given discipline or field, with assessments based where possible on the anonymity of the assessors. Peer review is also a part of the selection process in that assessors place a high weighting on the applicant's own publications, which are themselves published in peer-reviewed academic journals and peer-reviewed books and monographs.

Historically, the heart of the review process with individual researchers is the academic discipline. Disciplines imply independence, not only from government but from other academic disciplines as well. Thus NSERC has had core discipline–based panels dealing with core disciplines such as physics, chemistry, and engineering; SSHRC with panels on economics, political science, sociology, etc.; and the MRC-CIHR on biology, and the like.

However, disciplines change and new disciplines or aspiring disciplines emerge over time, as do calls for interdisciplinary research and thematic research whose intent is to draw upon various disciplines. Thus disciplines both represent the core of the granting process for individual applicants and allow science to progress and grow on the basis of a common but changing stock of theory and knowledge, and they provide the basis for its continuous evaluation. Academic disciplines also represent the academic equivalent of 'silos,' an often-lamented institutional feature that comes to light when people study complex organizations and bureaucracies. Some of the first evidence of aspiring new disciplines forming comes with the establishment of new or breakaway academic journals. The plethora of new journals is often followed by the later establishment of new panels for peer review within the granting councils and also possibly across the three granting bodies (see more below).

For example, the CIHR now has over a hundred peer review panels, reflecting not only changes among core disciplines but also the above-noted formation of thirteen virtual-reality institutes. NSERC now has twenty-six peer review panels, 40 per cent of which did not exist ten years ago. And, of course, as noted in our initial profile, NSERC was arguably the first of the three granting councils to embrace strategic research initiatives as a complement to its mainstream investigator-led research granting activity. SSHRC had also seen some growth in the number and nature of its peer review panels well before the recent consultation exercise aimed at transforming it from a granting council to a knowledge council.

When strategic or thematic research increases as a part of the granting body research portfolios, there are also some impacts on the nature of peer review. Peer review is still central to all granting decisions but it does change in subtle ways. In core investigator-driven research applications, an applicant's peers are assessing the proposal on the basis of core scientific criteria and of its quality and contribution to knowledge. In strategic research the applicant's peers must, in addition, assess whether the grant falls properly within the remit of the strategically defined field and is relevant and useful to that field. There is also increased pressure in such fields to have some form of community review or user review presence in the peer review assessment process (Shove and Rip 2003; Davenport et al. 2003). Community reviewers have been built into some of the CIHR's peer review processes, and this procedure has also led to regular review reports on the CIHR's peer review system and reforms (Canadian Institutes of Health Research 2005).

Accountability and Value for Money Spent on Research

The three granting bodies have been broadly influenced by, and have been practitioners of, varied kinds of accountability in the sense of both accountability 'to whom' and accountability 'for what,' and even accountability 'over what time frame.' In this section we look at these accountability relations quite broadly though a discussion of: a) changed funding mechanisms and related decision-making processes; b) the development of the New Public Management; and c) overall relationships of the granting bodies with universities as corporate entities regarding spending, research ethics, and intellectual property.

Overall changes in funding mechanisms have impacted the granting councils directly, but also indirectly through reverberations from the establishment of other funding bodies such as the Canada Foundation for Innovation (CFI), and the above-mentioned Canada Research Chairs program. The grant process has always been competitive at the individual investigator/applicant level, but often this occurs in a manner that falls below the radar screens of political attention. However, grant programs such as the Networks of Centres of Excellence (NCE), which the granting bodies jointly administer, brought more visibly to the surface *collectivities* of researchers – virtual reality networks that are geographically dispersed, but in practice also identified with certain regions or clusters of big versus small universities (Atkinson-Grosjean 2006).

This creates political concerns about accountability for the regional distribution of funds. This is all the more likely if some kinds of funding – for example, NCE funding – require applicants to 'bring money to get money.' Levered money and/or partnership funding, while not dominant in the overall granting body funding picture, has nonetheless become a growing part of it, especially in the last ten years, as has infrastructure funding provided by the CFI (see chapter 5).

When strategic or thematic research increases as a part of the granting body research portfolios, there are also some impacts on the nature of peer review, which in its own way is also a form of accountability. Peer review is still central to all granting decisions, but it does change in subtle ways. In core investigator-driven research applications, an applicant's peers are assessing the proposal on the basis of core scientific criteria and the quality and contribution of the proposed research to knowledge. In strategic research applications peers must also, as we have seen, assess whether the grant falls properly within the remit of the strategically defined field and is relevant and useful to that field.

There is also increased pressure in such fields to have some form of community or user review presence in the peer review assessment process. Community reviewers have been built into some of the CIHR's peer review processes, and this has led in turn to regular review reports on the CIHR's peer review system and reforms. To give one example, this procedure has been a part of SSHRC research grants dealing with Aboriginal peoples. The analysis of the CFI in chapter 5 also discusses differences between peer-review and the CFI's 'merit-based' review processes.

In addition, one needs to take account of the impacts of the other related funding programs created by the federal government, such as the previously noted CFI and Canada Research Chairs. As chapter 5 shows, the CFI funds university and hospital infrastructure proposals. Its multi-billion dollar funding is of the levered kind and is separate from granting council funding. Nonetheless, its funding can create future demands for normal research grant funding. Once new buildings, labs, or equipment for the research task have been created, research personnel and operational funding for researchers (faculty and students) who will use them must also be found. In addition, the CFI has required universities to submit research plans as an accountability mechanism, and this also feeds back in various ways into the individual and strategic research grant processes.

Finally, in this discussion of funding accountability, it is useful to take note of the issue of research overheads. Canadian research grants do not fund university or hospital overhead costs. In other countries such as the U.K., the research payment funds a further add-on of 25 to 40 per cent of the costs of the grant. The historical assumption in Canada has been that the provinces cover this key expense, but this is not a reliable assumption to make. Canada's R&D system has therefore suffered from a major competitive funding gap that is very real and that effects overall research capacity in a serious way.

Because of this gap, Canada's universities individually and, as chapter 3 has shown, through the work of the AUCC, have lobbied the federal government to provide some overhead funding that would be dispersed separately to universities through granting council mechanisms in proportion to the grants that individual universities receive. While both the Liberal government and the current Conservative government have increased overhead funding, Canada's researchers still face a formidable obstacle and are very much playing a catch-up game.

The second changed accountability value and relationship is more broadly institutional and managerial. As noted in chapter 1, it has centred on changes in public administration that focused initially on the so-called New Public Management (NPM), and then also on changes in and concerns about accountability 'to whom.' Broadly speaking, the NPM concepts were important because they sought to have federal agencies of all kinds (including the granting bodies) become less hierarchical so as to become more focused on clients, customers, and service delivery, and to become more networked and engaged with partnerships and with the above-noted more diverse ways of shared and levered funding. There were also more formal requirements to show management by results. We have already seen glimpses of these changes in earlier sections.

These changes in part made core concepts of accountability and value for money 'up' to ministers more complicated because in some respects NPM implied accountability 'down and out' to clients and partners and also 'across' to other federal partners and possible co-funders. But traditional hierarchical accountability has also taken on a renewed focus with the Harper Conservative federal government because this government won power in 2006 following an election that had been fought over core issues of corruption and misuse of public funds and of Ottawa being perceived as having a mentality of entitlement. Thus, the three granting councils are subject to any number of potential policy- or management-induced pressures and signals, which they have to interpret, act on, or resist depending on the impacts they think they will have on their core statutory mandates.

In a similar vein, it is useful in a historical sense to take note of the overall *accountability relationships of the granting bodies with universities as corporate entities regarding spending, research ethics, and intellectual property*. These can only be noted briefly but they are important. At their core, research grants go to individuals or small teams of researchers, but the funds must be administered by the faculty member's home university to ensure proper use of and accountability for those funds. As regular grants and thematic program funding increased, this accountability and management-centred task became more complex and costly. Along with other factors, it has led as well to the previously mentioned university concerns about their overhead costs and the fact that grants do not cover overheads.

Eventually, Canada's universities developed an overall accountability framework titled the *Memorandum of Understanding on Roles and Responsibilities* (NSERC 2002). This is an agreement with the federal government developed through the work of the Association of Universities and Colleges of Canada (AUCC) with regard to the management of funds, policies, and expectations in agency-funded research and the assurance of due diligence. In addition, other key facets of how research is conducted had to be subject to cross-university rules. One of these was research ethics, where in 1998 an AUCC and federal effort led to a Tri-Council Statement on Ethics (TCPS). As Levasseur's analysis in chapter 10 shows, it provides policy guidelines that determine conditions of eligibility for federal grants from the three bodies. The granting bodies are careful not to cast these as regulations in a formal legal sense, but universities adhere to them de facto and grant-holding faculty and students are subject to them within universities. As Levasseur's chapter shows, ethics rules have created some cross-granting council concerns and differences because some in the natural sciences and social sciences believe that the rules are too heavily dominated by a biomedical model skewed in favour of the much greater necessity for controls in research on human subjects, the detailed focus of the rules being much more appropriate to health research but present in other granting realms as well (SSHRC 2006).

A further related area where there have been impacts on university practice but not necessarily common rules is that of intellectual property (IP). Federal S&T and innovation policy have sought to have publicly funded research lead to demonstrable returns and evidence of innovation and commercialization, first through patents, and then through spin-off companies and actual product development and sale in markets. Thus, gradually universities and their varied offices of technology transfer and development have developed internal policies with regard to the handling of IP and where ownership of such IP resides vis-à-vis the university, faculty, and often graduate students.

The Federal Government and Canadian Society's Need for Objective and Useful Social, Economic, and Policy-Relevant Research

While the discussion above of the three granting agencies' core business of investigator-led peer-reviewed research remains true overall, there is little doubt that a key change in the federal government–university relationship in the last twenty years has been continuous pressure for more thematic or strategic research. There is also, as chapter 1 has shown, more pressure for the dissemination of publicly funded research to serve Canadian society and business and commercial needs and goals. The nature of this value and the relationships it creates and changes are examined in three ways. First, the nature of the federal government's own policy-relevant research needs to be understood. Second, the pressures overall for more thematic or strategic research are traced. And third, this value in terms of the role of granting bodies as supporters of the development of highly qualified personnel (HQP) is highlighted.

Federal Government needs for policy-relevant research is a key overall value and norm in the federal government-granting council relationship. To discuss this element of core federal granting body values and relations, we need to examine, first, how views and dynamics have evolved, and then the different dynamics among the three granting bodies and their interactions with the federal government as a complex departmental entity.

The overall view of federal departments and their changing policy-research needs is difficult to discern. More particular views, links, and interactions will emerge below regarding each granting agency's particular policy interactions, but an overall federal departmental view involves about thirty-five departments. We have not attempted to canvas these views, but one interesting discussion in 2005 does give us some useful glimpses.

The 2005 discussion process involved quite a broad conversation between sixteen senior bureaucrats and senior officials at SSHRC, which occurred in the wake of SSHRC's own endeavour, described above, to transform itself from a granting council to a knowledge council. The discussions had been preceded by a preliminary position paper that had scoped out some key issues regarding the linking of academic research and policy development. Among the issues and points (contextual and substantive) raised and discussed were the following:

- Given our current state of knowledge, there are no right or wrong answers in terms of how best to promote linkages between policy and research, just choices regarding experimentation.
- It is important to bear in mind that research serves multiple ends, informing not just policy development but other policy-related activities and persons connected to the policy process as well.
- Though research and policy operate along different timelines, the fact remains that most fundamental policy questions are durable. As well, there tends to be agreement between the academic and policy communities regarding broad research and policy priorities.
- It is appropriate for SSHRC to be an agent of change but this role will need to be implemented carefully, maintaining balance across portfolios. In adopting new directions, SSHRC will need to concentrate on those areas where it enjoys a strategic advantage relative to other policy-research stakeholders. One of SSHRC's main advantages resides in its ability to manage incentives through funding allocations. SSHRC's spending power affords the Council a powerful means for influencing university and researcher priorities and signalling the need for a change in direction. SSHRC is not presently equipped to dispense policy advice.
- Universities are not just in the business of training future academics. They will need to be enlisted in SSHRC's efforts to engage researchers in policy work and knowledge transfer activities. Universities have an obligation to equip students and faculty to contribute to socially desirable pursuits, such as policy development.

- By emphasizing knowledge mobilization and knowledge transfer, SSHRC is seeking to leverage federal interest and investment, especially by departments that face horizontal issues. In this regard, SSHRC's transformation can serve as a catalyst. Ultimately, however, the *lead* for improving receptor capacity rests with government. So too does the challenge of responding to and taking advantage of SSHRC new directions.
- For the policy community to be persuaded of the value of research, it will need to benefit from it. This creates a shared interest in transferring knowledge. Experience has shown that sustained involvement with research and researchers has a favourable effect on policy valuations of research.
- The benefits that accrue from linking policymakers and researchers are not strictly linear. Research benefits not only policy but also policy*makers*. In addition to yielding information that aids in policy decisions, participation in research activities produces different ways of thinking and reflection – in short, learning. Government has been exclusively preoccupied with managerial training to the detriment of scientific learning and policy development capacity (Burstein 2005, 2–3).

There appeared to be a consensus among the participants about the implications of some of these concerns, although the last sentence quoted above is a highly debatable assertion. One implication was that a granting body like SSHRC still had to do both investigator-driven and targeted strategic research. Another was that the academic reward and incentive system was not conducive to young academics being engaged in policy research and in the dissemination of policy research to policy practitioners. For this and other reasons there was a sense that university research has a limited value for public policy and policymakers and that policy think tanks may be seen as more relevant. A further implication that officials acknowledged was important was that the government also had a limited 'receptor capacity' and that university research was undervalued. A final implication was that multidisciplinary research was needed because so many new but permanent enduring issues were horizontal and multi-departmental and multi-field in nature (Burstein 2005, 3–4).

These views are in some respects not surprising but also have to be seen against a larger set of developments that go beyond the social sciences realm of the SSHRC. Other recent commentaries by federal officials in the health and general science fields reinforce the above mixture of negative and positive views about academic science and public policy as 'strangers in the night' (Sexsmith 2006; Carty 2006; Edwards 2004).

One essential backdrop historically was that the many policy units of federal departments had to absorb significant budget and personnel cuts during Program Review in the 1995 to 1997 period and beyond. This meant a definite loss of policy advice capacity and also of policy research per se. Some individual policy units have since had their funding increased. Nonetheless, in the late 1990s, the federal response to the earlier cuts was partly to rebuild by seeking to pool resources and by networking among its policy shops, but also by engaging with think tanks and university policy researchers. This was undertaken partly through the role of the Policy Research Initiative (PRI) located in the Privy Council Office (but recently relocated to the Human Resources and Skills Development Canada), but anchored by a deputy-minister level committee.

Both of these developments suggest that the federal government has still not fully rebuilt its policy capacity, nor is it easy for it to do so. It is difficult for policy shops and regulatory analysis units ever to have on staff an optimum range of expertise in a fast-changing world. Policy and related analytical shops or units face formidable choices of 'making or buying,' and also of renting and borrowing policy researchers/advisors and substantive research. Hence the interest in how to mobilize the SSHRC-supported social science community.

One point worth noting about the social science community and senior federal officials is that the latter are likely to have a more natural affinity with the social sciences overall than with the natural or health sciences. By this I mean that very few senior officials at the DM and even ADM level have science backgrounds, and hence the knowledge and appreciation gap is greater for the research that might come in from granting-based research from NSERC and CIHR. There are, of course, many other factors involved in these kinds of 'receptor' issues, including technical people who are the interpreters of published research but who are also often people who are summarizing other people's summaries of the policy implications of someone's research or a team of researchers' findings. So the policy supply chain is not just a supply issue but also an issue of interpretation and communication by intermediary staff both inside and outside governments.

As one goes analytically from the federal–social science links to the research realms encompassed by the natural and health sciences one encounters a typically even more complex set of dynamics regarding just what kinds of policy research federal departments might need and what the supply–interpretation chain is. Policy research may or may not include the kind of research needed to support varied regulatory and monitoring tasks, much of which is classified within the federal government, not as R&D, but as related science activities (RSA).

The full set of needs also encompasses the overall roles of federal science-based labs, agencies, and units. These key elements of public-interest science have also been going through their own version of budget cuts in the mid-1990s and beyond, as well as debates about whether government science should be 'in-house' at all or needs to be transparently and structurally arm's length (Doern and Kinder 2007; Doern and Reed 2000). The federal government has not fully replenished or reshaped these kinds of in-house R&D and S&T needs, largely because its own S&T and innovation priorities have focused on getting resources out to universities and to the business community. Government science overall, as chapter 1 and the Appendix show, has been in a decidedly third-place position, which, in a three-way horse race, means that it is dead last.

Within this larger backdrop of federal need for policy research, we now look at the different dynamics regarding the three granting bodies' relations with the federal government. Not surprisingly, the nature and dynamics of the federal government's interactions with the granting councils varies significantly, and also takes place in different arenas of discussion and pressure. It is obvious that such relationships are not the only ones that both parties to the relationship foster. The granting bodies have their own links with relevant clientele and stakeholders, including academics, the universities both directly and through their national associations such as the AUCC and CAUT, business interests, and social stakeholder groups. The federal government is of course also pursuing multiple channels of influence and discussion in addition to those which it has directly with the granting councils.

Like the other two agencies, SSHRC sits on the committee of S&T Assistant Deputy Ministers (ADMs), which meets monthly, and through which government-wide S&T issues are made known to the councils and council concerns are expressed to senior federal officials. In the past there have also been links through the Council of Science and Technology Advisors (CSTA), an external advisory committee to the

Cabinet on matters dealing with overall federal S&T policies and operations. However, the CSTA has been eliminated by the Harper Conservatives in favour of a consolidated single S&T advisory body. Additionally, there are some potential arenas of exchange through central policy shops such as the previously mentioned Policy Research Initiative (PRI), though these are only one internal arena or access point for the granting councils and the researchers they fund.

SSHRC operates mainly in a responsive mode with its core academic community but, as noted above, it has sought out, and been subject to more pressures and networking on, thematic and strategic research. The initiative on the New Economy, which SSHRC administers, arose through joint discussion, whereas a research program on the Social Economy came more or less 'out of the blue,' and was handed to the Council for the administration of its research grants. As chapter 1 has shown, the Harper Conservative S&T strategy also handed new commercialization programs to SSHRC.

In recognition of its aspirations to be more a knowledge council rather than just a granting council, SSHRC has held its own preliminary competition among social scientists for initial suggestions for thematic research teams and topics. The response was far greater than SSHRC expected, and thus shows that teams of researchers will follow these funding signals or at least test them out. SSHRC has since responded to this by announcing a clusters competition for research communication and networks in strategic subject areas. Up to seven clusters will be funded, with grants of up to $300,000 per year for up to seven years. As a result of these changes, SSHRC also now has a Vice-President for Partnerships and a Vice-President for Programs, with the former seeking to forge greater links with non-academic stakeholders, including the federal government.

SSHRC and NSERC, as we have seen, are both part of the Industry Canada portfolio but report directly to the Minister of Industry, through whom they report to Parliament. Both granting bodies and the Minister are aware of the legal independence of the granting bodies and of the need to keep the granting process free from political interference. Meetings with the Minister are very infrequent, but when they occur they become a conduit for any joint concerns or ideas about the granting bodies and/or about what is on the Minister's mind or agenda.

NSERC's relationship with the federal government is the product of a different history and pattern of linkages. NSERC gets somewhat more attention in its Industry Canada portfolio relations, for the simple reason that it is natural sciences and engineering and related S&T that are most closely linked to the larger Industry Canada industrial and innovation mandate and to its lead S&T policy role. Industry Canada does not typically see SSHRC or the social sciences as being 'top of mind' in its core industrial/commercialization mandate.

NSERC functions as well through the advice provided not only by its council members from across Canada but also through two advisory committees, one on research and another on research partnerships. It has been quite welcoming of joint funding research areas and over the years has worked with Natural Resources Canada and Agriculture and Agri-food Canada, to mention only two departments. It has also been involved, through its links with the ADM Committee on Climate Change, with climate change and environmental technologies funding activity. It has fashioned as well a quite extensive set of research funding relations through joint university–industry research. Through various research granting mechanisms it now has links with over 1300 companies.

NSERC has looked closely at the CIHR model of virtual reality Institutes, but concluded rather quickly that this was not a model it wanted to emulate. The conclusion was that there were no obvious divisions or themes, nor was there much latitude for new approaches to integration. This does not mean that NSERC's realms of science and engineering were not subject to pressures for thematic or strategic research expressed without having to resort to an institute model. As noted earlier, NSERC's peer review panel structure has changed as new panels were added. Its university researchers were also extremely successful in the various NCE competitions, which were thematic and which were in some limited respects a model for the later CIHR.

The CIHR's links with its minister, the Minister of Health, and through the Health Canada portfolio of agencies is quite different from that of SSHRC and NSERC vis-à-vis Industry Canada. The latter granting bodies are unlikely to be actively sought out for S&T policy advice per se by Industry Canada, but CIHR and its Institutes are sought out quite actively by Health Canada for advice on heath research policy and on substantive health policy issues such as HIV Aids, SARS, and pandemics. CIHR does have links with the government-wide ADM S&T Committee, but also with the ADMs involved in the federal biotechnology strategy.

Both through the CIHR Institutes and through the CIHR's overall planning and priorities process, numerous other thematic and strategic areas are constantly being brought forward. These have recently included research issues related to global health, the biology of aging, and nano-health research. There is also an interaction here with the CIHR peer review panel structure. As noted earlier, the CIHR has over one hundred panels, compared to about forty under the previous Medical Research Council system. New panels are sometimes needed simply for workload reasons, but more broadly they are driven by both bottom-up and top-down pressures and demands for new realms and groupings of research. The CIHR prides itself on being activist about these developments, knowing that health research in the present era is highly networked and composed of virtually self-assembled teams of researchers practising interdisciplinary and discipline-based research.

The *pressures for thematic or strategic research* increasingly being applied on the granting bodies as 'boundary organizations' are accordingly numerous and varied. They come, as stressed above, from the federal government via its overall S&T and innovation policies and also from particular federal departments, which lobby within the federal structure for targeted funding of thematic areas linked to their particular departmental mandates and policy concerns. Pressure also comes from business and NGOs, either directly on the granting council president or governing board members or indirectly via their lobbying of federal departments, which in turn lobby the granting councils. Last but certainly not least, pressures come from academics themselves, either entrepreneurial individual scholars or new emerging communities of academics, some of whose interdisciplinary lobbying was noted in the discussion above about the changing nature of disciplines versus thematic areas of study.

The nature and timing of these kinds of shifts towards greater thematic or strategic research varies among the three granting agencies. Arguably the shift started with NSERC. As noted earlier, from 1978 on, NSERC developed more strategic realms of research funding. NSERC also had a somewhat more natural affinity for, and linkages with, business and market-related research. This typically did not relate to explicit commercialization-focused research (see more below), but the natural sciences and engineering area had faculty and graduate students much more likely to gravitate towards industrial research and also post-university employment in the private sector. The development in the late 1980s and early 1990s of the Networks of Centres of Excellence (NCE) program, with strong links to NSERC (much more so than SSHRC), also acted as a spur to strategic research and to virtual research communities (Atkinson-Grosjean 2006). Such Centres of Excellence have become a global phenomenon, raising difficult issues as to how to evaluate their work and progress as networked realms of research (Lukkonen et al. 2006).

The CIHR is without doubt the main example of a granting agency which was, in effect, born in 2000 to demonstrate, in one fell swoop, the transformation towards a greater focus on strategic and applied targeted research and also multi-disciplinary research. A reading of its published reports and website material indicates that the thirteen new institutes tend to receive greater emphasis than its former MRC core investigator-driven research. But this is misleading in two important senses. As Joan Murphy's research points out, there are two key features to the larger origins of health research and the evolution of MRC and then CIHR (Murphy 2007). The first is that health research in Canada has always been more networked and thematic because the core of the research community did not reside just in universities but also in hospitals and charitable organizations. Indeed, only later did universities become a larger dominant player in health research. The second is that the CIHR is now, in some respects, two interacting organizations within one. Housed in its inner core is the older MRC core investigator-driven granting activity. And on its outer circumference, but growing larger, are the thirteen institutes. As noted earlier, they receive about 30 per cent of CIHR funding (several times more than strategic research did before 2000) but still leave 70 per cent for the agency's core investigator-led research grants. As pointed out in our initial CIHR profile, there is considerable overlap in these two kinds of research and in the nominal budgets described for each of them.

As noted earlier, SSHRC was arguably the last granting body to explicitly move in the direction of more strategic research. In part this was genuinely because the social sciences and particularly the humanities had historically been less used to working in thematic ways and their research was more inherently individual in nature and generally lower in cost. But the social sciences, too, were moving more to team research in the 1990s. SSHRC was also realizing that the social sciences had not done well in NCE competitions. It was also concerned that its research clientele (faculty and students) was by far the largest part of the total university graduate student population but, as we have seen, was not getting anything proportionate to that in resources and funding. Moreover, SSHRC had witnessed in 1999–2000 how the CIHR's transformation and introduction of virtual Institutes had, along with vigorous leadership and ministerial support, generated double the funding for health research. Thus SSHRC's effort to transform itself from a granting council to a knowledge council was part of a larger effort to position itself better in the shifting configurations of core investigator-driven versus strategic research.

The overall value of the social and economic use of research is also closely tied to granting agency *roles in encouraging and funding the development of highly qualified personnel* (HQP). The supply of HQP does not just happen. It must be nurtured and funded so as to meet overall HQP needs for Canada's economy and society and also in particular realms of research or S&T education already established, or in fast-forming new ones (nanotechnology, human genome, etc.). HQP goals through granting activity involve both university or hospital faculty, but also doctoral and Masters-level graduate students and even undergraduate students in certain fields or areas (e.g., women in science, entrepreneurship, and commercialization experience). Increasingly, they can involve programs to attract and retain foreign students who study in Canada but who possess key basic and applied research skills and ideas.

Demographics are also a large part of the HQP function in that, at the faculty level, significant swaths of university faculty will be retiring soon and have to be replaced. In addition, student populations also come in bursts not only in relation to normal demographics but also because of individuals returning to school or responding to the clarion call for lifelong learning. Last but not least, HQP issues can, in general or in particular fields, become caught up in the 'brain drain' and/or 'brain gain' dynamics caused by the greater mobility of research labour globally and, for Canada, vis-à-vis the dynamic U.S. market in particular. For example, the Canada Research Chairs program established in 2000 was seen in part as a federal response to real or imagined brain drain issues. Moreover, once the hundreds of Chairs were awarded it created an increased supply of net new research grant applications. As we have seen in our quick portrait of each council, each of the three granting bodies have had HQP issues firmly in mind for the last three decades. But HQP issues of demand and supply are also notoriously difficult to forecast, and even if one had perfect forecasts, budgetary and funding resources are by definition limited, with the result that, to some extent at least, the three granting bodies are competing with each other. Nonetheless, the need for HQP can only increase in a knowledge-based economy, and thus a key rationale for federal granting support is to ensure that Canada is creating a proper and ever greater supply of such key personnel to meet both economic and societal needs.

Conclusions

The purpose of this chapter has been to examine core values in the interactions between the federal government and universities in the research granting process through a comparative analysis of the evolution of NSERC, the CIHR, and SSHRC, and through an examination of recent changes in federal policy concerns and capacities.

The analysis has shown that federal needs for research have been influenced by a variety of forces and ideas, including budget cuts to internal departmental policy units, increased funding in some but not all research areas, pressures for dealing with ever more complex horizontal policy and governance problems and challenges, weaknesses in the federal government's own so-called receptor capacities and knowledge, and also perceptions that most research and most researchers do not necessarily produce research that is immediately or easily policy-relevant. Finally, it needs to be emphasized that the federal government has some thirty-five ministerial departments (and dozens more arm's-length agencies) and therefore numerous points of access for such research, whereas the granting bodies have formal relationships with only two, Industry Canada and Health Canada, and informal relationships with several others but by no means more than a small subset of them.

The chapter has shown the features and changing cultures of the three granting bodies and their constituent academic research disciplines. Each of the three agencies has been influenced by the need to reconfigure the proportion of its budget that goes to individual investigator-driven research as opposed to thematic and strategic research, and has to adapt to and utilize new research technologies, develop policies and practices on research ethics and intellectual property, and respond to increased demands for basic accountability and greater dissemination of publicly funded research to various users of such research.

Overall, the patterns of change in the federal government–university relationship through the granting process have been complex and nuanced but also significant. Clearly, there is close linkage and reinforcement between the first two values examined, namely, the relative independence of researchers from government and the importance of peer review and the peer review process. The first is valued for its own sake as being central to open research, and the latter is valued as central to the nature of how knowledge is structured and reviewed by others with expertise and knowledge.

While the focus in this chapter has been on relationships between the federal government's granting councils and universities in an overall sense, there are further existing relationships between universities and academic researchers that are also being changed. There are pressures on individual researchers for more collaborative research among individual researchers and across disciplinary lines. There are also entrepreneurial academic researchers who are among the advocates of more thematic research and targeted research. Individual researchers may have strong proclivities to do research that is more relevant to society and the economy and to communicate their research to various communities other than those reached by peer-reviewed publications.

The third and fourth values examined also have close compatibility in that taxpayers' money and how it is used is central to democratic governance and therefore research funded by such taxpayer funding is also subject to clear norms of accountability and value for money. No one would dispute this ultimate need and the value which underpins it. Similarly, it is not in the least abnormal that the federal government as the main research funder would also want to have that research feed into the policy problems and challenges of the day and of the future and into dealing with the needs of other interests and communities in the economy and society.

What, then, of possible integration or tension between the two conjoined pairs of values? In these interacting realms, some tension is highly likely as both of the players – the federal government and university-based academics – defend their often differently ranked values in the research granting process. This tension, but also the interdependence involved, exists in other realms as well, such as the state versus the press or media, and the state versus markets. But in the granting process, there are elements of some integration and understanding across these full value terrains as well. Within a certain band of acceptability the federal government at both the ministerial and official level does recognize the need for research independence and for peer review values and processes. Academics recognize the need for, and value of, accountability regimes, and for research to address economic and social problems and to contribute wide-ranging solutions and improvements. Some differences undoubtedly exist between the natural and social sciences in this regard, but on the whole there appears to be joint recognition at the level of overall values that some broad form of integration is possible.

However, boundary problems and tensions remain. Indeed, we have noted analytically at several points the growing importance of boundary organizations and of boundary processes. The granting bodies themselves are boundary organizations, as are universities themselves in a wider sense. Boundary processes also occur through peer review committees, which change over time as new realms of science, or, equally, new thematic areas of economic and social concern, wend their way in and amongst the interstices of established peer communities and journals, and newly forming ones.

A final important theme which emerges from the chapter as a whole is the fact that many of the changes in this relationship have occurred well below the normal radar screens of both the federal government and universities. This suggests, in the author's view, that the granting councils, the federal government, and the universities need some broader, open, and regular annual forum for discussion and review that both defends and examines these values and that assesses the boundary organizations and boundary processes that underpin them not only in a knowledge economy but also in a knowledge society.

REFERENCES

Atkinson-Grosjean, Janet. (2006). *Public Science, Private Interests: Culture and Commerce in In Canada's Networks of Centres of Excellence*. Toronto: University of Toronto Press.

Beesley, Anne E., Paul Cunningham, and Luke Georghiou. (1998). 'Convergence of Research: The Government, Public and Independent Sectors.' In *Science and Technology in the United Kingdom*, 2nd ed., ed. Paul Cunningham, 127–58. London: Cartermill International.

Burstein, Meyer. (2005). 'Summary of April 5th Meeting.' Report for the Social Sciences and Humanities Research Council of Canada, April 13.

Canada. (1996). *Science and Technology for the New Century: A Federal Strategy*. Ottawa: Industry Canada.

– (1999a). *Public Investments in University Research: Reaping the Benefits*. Ottawa: Industry Canada.

– (1999b). *A New Approach to Health Research for the 21st Century: The Canadian Institutes of Health Research*. Ottawa: Canadian Institutes of Health Research.

– (2005). *Canadian Institutes of Health Research, 2005–2006 Report on Plans and Priorities*. Ottawa: Treasury Board Secretariat.

CIHR. (2005). *First Report on Peer Review Innovations*. Canadian Institutes of Health Research, June.

- (2006a). *Research Funding Overview*. Canadian Institutes of Health Research.
- (2006b). *Response to Budget 2006–CIHR Budget Decisions*. Canadian Institutes of Health Research.
Carty, Arthur J. (2006). 'Towards a Framework for Major Science Investments in Canada-Social and Health Science Considerations and Implications.' Presentation to Conference on Longitudinal Social and Health Surveys in an International Perspective, Montreal, January 26).
Davenport, Sally, Shirley Leitch, and Arie Rip. (2003). 'The "User" in Research Funding Negotiation Processes.' *Science and Public Policy* 30 (4): 239–50.
de la Mothe, John. (2003). 'Ottawa's Imaginary Innovation Strategy: Progress or Drift?' In *How Ottawa Spends, 2003–2004: Regime Change and Policy Shift*, ed. G. Bruce Doern, 172–86. Don Mills, Ont.: Oxford University Press.
Department of Finance. (2006). *The Budget Plan 2006: Focusing on Priorities*. Ottawa: Department of Finance.
Doern, G. Bruce. (1996). 'Looking for the Core: Industry Canada and Program Review.' In *How Ottawa Spends, 1996–97: Life under the Knife*, ed. Gene Swimmer, 73–98. Ottawa: Carleton University Press.
Doern, G. Bruce, ed. (2006). *Innovation, Science and Environment, 2006–2007: Policies and Performance*. Chapter 1, 3–34. Montreal: McGill-Queens University Press.
Doern, G. Bruce, and Jeff Kinder. (2007). *Strategic Science in the Public Interest: Canada's Government Laboratories and Science-Based Agencies*. Toronto: University of Toronto Press.
Doern, G. Bruce, and Richard Levesque. (2002). *NRC in the Innovation Policy Era: Changing Hierarchies, Networks and Markets*. Toronto: University of Toronto Press.
Doern, G. Bruce, and Ted Reed, eds. (2000). *Risky Business: Canada's Changing Science-Based Policy and Regulatory Regime*. Toronto: University of Toronto Press.
Edwards, Meredith. (2004). *Social Science Research and Public Policy: Narrowing the Divide*. Academy of the Social Sciences, Canberra, Australia.
Kinder, Jeff. (2003). 'The Doubling of Government Science and Canada's Innovation Strategy.' In *How Ottawa Spends, 2003–2004: Regime Change and Policy Shift*, ed. G. Bruce Doern, 204–20. Don Mills, Ont.: Oxford University Press.
LaPointe, Russell. (2006). 'The Social Sciences and Humanities Research Council: From a Granting Council to a Knowledge Council.' In *Innovation, Science, Environment, 2006–2007: Canadian Policies and Performance*, ed. G. Bruce Doern, 127–48. Montreal: McGill-Queen's University Press.
Lopreite, Débora. (2006). 'The Natural Sciences and Engineering Research Council as a Granting and Competitiveness-Innovation Body.' In *Innovation,*

Science, Environment, 2006–2007: Canadian Policies and Performance, ed. G. Bruce Doern, 105–26. Montreal: McGill-Queen's University Press.

Luukkonen, Terttu, Maria Nedeva, and Remi Barre. (2006). 'Understanding the Dynamics of Networks of Excellence.' *Science and Public Policy* 23 (4): 239–52.

Miller, J.R. (2006). 'A Short History of SSHRC.' Social Sciences and Humanities Research Council, May.

Murphy, Joan. (2007). 'Transforming Health Sciences Research: From the Medical Research Council to the Canadian Institutes of Health Research.' In *Innovation, Science, Environment: Canadian Policies and Performance, 2007–2008*, ed. G. Bruce Doern, 240–61. Montreal: McGill-Queen's University Press.

NSERC. (1999). *Ten Years to 2000: A Strategic Document*. Natural Science and Engineering Research Council.

– (2002). 'Memorandum of Understanding (MOU) on the Roles and Responsibilities in the Management of Federal Grants and Awards.' Natural Science and Engineering Research Council. Available from: http://www.nserc.gc.ca/institution/mou_e.htm (cited 27 November 2008).

– (2003). *Investing in People: An Action Plan*. Natural Science and Engineering Research Council.

– (2004a). *New Government for Canada, New Vision for NSERC*. Natural Science and Engineering Research Council.

– (2004b). *Update on NSERC's Vision Initiatives*. Natural Science and Engineering Research Council.

– (2006). 'About NSERC: Facts and Figures.' Natural Science and Engineering Research Council.

Office of the National Science Advisor. (2005). 'A Framework for the Evaluation, Funding And Oversight of Canadian Major Science Investments: Draft Discussion Paper.' 31 January.

Sexsmith, Wendy. (2006). 'Strangers in the Night?: Exploring Opportunities for Collaboration between Government and Academia.' Presentation to Health Regulation Conference, Ottawa, 23 March.

Shove, Elizabeth, and Arie Rip. (2000). 'Users and Unicorns: A Discussion of Mythical Beasts in Interactive Science.' *Science and Public Policy* 27 (3): 175–82.

SSHRC. (1979). *1978–79 First Annual Report*. Social Sciences and Humanities Research Council.

– (2004). *From Granting Council to Knowledge Council: Renewing the Social Science and Humanities in Canada*, vol. 2. Social Sciences and Humanities Research Council, January.

– (2005). 'The National Discussion on Protection of Human Subjects in Research.' Social Sciences and Humanities Research Council.

– (2006). 'About SSHRC.' Social Sciences and Humanities Research Council.
Treasury Board of Canada Secretariat. (2005). *Report on Plans and Priorities, 2005–2006: Natural Sciences and Engineering Research Council of Canada.* Treasury Board of Canada Secretariat.

5 The Canada Foundation for Innovation as Patron and Regulator

DÉBORA LOPREITE AND JOAN MURPHY

The aim of this chapter is to provide an analysis of a relatively new independent agency, the Canada Foundation for Innovation (CFI), created by the government of Canada in 1997 to support research infrastructure.[1] Such university infrastructure had not been supported before in any focused way by the federal government. Its research grant funding through the three granting councils examined in chapter 4 has focused on the operational costs of research, with even overhead costs only being supported in very recent years. We examine the CFI in the context of the Canadian university system. The CFI also supports research hospitals, but these are not examined in this analysis.

The CFI was given an initial budget of $800 million, with large later increases coming in other federal budgets in the late 1990s and beyond. The sponsoring Minister of Finance, Paul Martin, recognized that innovation in the knowledge-based economy was a competitive reality and Canada could not afford to be left behind. The Harper Conservative government has given it a further $510 million in investment funding.

Late in 2006, the President of the CFI reported to a Parliamentary Committee that the CFI had, since 1997, committed $3 billion in 4,700 esearch infrastructure projects at 128 institutions in 62 municipalities across the country (Phillipson 2006, 1). Moreover, by 2010, the total capital investment in research infrastructure by the CFI, the research

---

1 Research for this chapter is based mainly on the reports and studies cited but complemented by a few confidential 'not for attribution' interviews with senior CFI officials. The three illustrative looks at particular universities involved the selection of universities based on their size (two small and one large), and also some very limited regional coverage.

institutions, and their partners, will exceed $11 billion (Phillipson 2006, 2). As part of its funding agreement with the federal government, the CFI is committed to supporting several national objectives, including 'economic growth and job creation, as well as health and environmental quality through innovation' (CFI 2006, 3). These commitments also include recent Harper Conservative government priorities announced in its 2007 S&T strategy already previewed in chapter 1 (CFI 2007).

We examine the CFI as a *patron* based on the CFI's role as a funder and promoter of research and development infrastructure particularly in universities. Although the CFI is an arm's-length foundation and is independent of politics in the sense of partisan politics, it does have to deal, as all patrons do, with small-p politics among universities in the form of distributional disbursements regionally or among the different realms or disciplines of science and the infrastructure that each needs. These are, of course, not part of the CFI's terms of reference, but they are a part of how diverse research interests lobby the CFI and seek to change its program structure.

In addition we argue that the CFI acts as a *regulator*, as defined and discussed in chapter 1, because investments in research and development infrastructure must conform to certain 'rules' set by the CFI, including requirements that universities submit strategic research plans and also that funding be levered. In short, universities do not get infrastructure money unless they bring money in the form of funding commitments from other funders, public or private. Universities have widely differing capacities to obtain these levered money commitments, which constitute 60 per cent of all CFI awards. Aspects of regulation also occur in the realm of defining what is eligible for funding, and hence in defining what in fact 'infrastructure' is as capital spending rather than operational spending (CFI Eligibility). Such issues also inherently arise in the CFI's role of developing research 'capacity,' in determining what 'innovation' means in these funding choices, and also in determining what kinds of research capacity-building might occur when CFI funds are allocated. Capacity-building involves both complex physical-technical capacity and human capacity in the form of Highly Qualified Manpower (HQM) and managerial and even entrepreneurial capacity. Innovation, as we have seen in chapter 1, can also mean a broad sense of research support in the full spectrum of R&D, or in more precise terms, the development of new products and production processes.

This chapter is organized as follows: First, we provide an overview of the CFI's structure, mandate, policies and programs, changes in program architecture, and how programs have been formally evaluated. Second, in order to capture some of the impacts of the CFI in the Canadian university system we present brief illustrative looks at two smaller Canadian universities (Saint Mary's and Brock) and one large university (the University of Toronto) to illustrate the patron and regulating role of the CFI in contributing to and, in some cases, reformulating research in Canadian universities.

With these early mandate and program issues developed, the chapter then links the CFI back to overall federal research and innovation policy, as introduced in chapter 1, and to related issues concerning the knowledge-based economy (KBE) and innovation, the accountability of foundations as research boundary organizations, the New Public Management as a linked governance reform idea and set of practices, and differences between peer review versus 'merit review' as introduced in chapter 1. Conclusions then follow.

The CFI: Structure, Mandate, and Core Programs

Structure

As chapter 1 has already stressed, the CFI structure is of more than usual interest because, when it was formed as a foundation, the foundation model was itself a new form of federal governance structure. As a newer form of what chapter 1 has referred to as a 'boundary organization,' it has subsequently been criticized amidst concerns about accountability by the Auditor General of Canada and others (CFI 2006; Aucoin 2003, 2005). Because of the CFI's creation as foundation in 1997, the federal government appoints only a minority of the board of directors. The board of directors meets three to four times a year. An annual public meeting is also held each year and is widely publicized in several of Canada's leading newspapers. The Board of Directors consists of fifteen members, with seven, from a variety of research and academic backgrounds, appointed by the Canadian Government. The Directors are appointed for a three-year renewable period. They must demonstrate previous expertise in the private, institutional, academic, research, and government sectors. One Director on the Board is a representative from one of the federal granting agencies.

The Board of Directors reports to Members – a higher governing body similar to a company's shareholders, but representing the Canadian public. These Members are responsible for the appointment of eight of the fifteen Directors. They receive audited financial statements, appoint auditors, and approve the Annual Report at their annual meeting.

The CFI's Board of Directors makes final decisions on projects to be funded and sets strategic objectives in the context of the funding agreement the CFI has with the federal government. It approves annual plans and objectives, and reviews the outcomes of these objectives each year. It regularly reviews issues from a risk-assessment perspective, determining what risks are acceptable and ensuring that appropriate mitigation steps are in place. As well, the Board sets the CFI's overall compensation policy, and specifically sets compensation for management (CFI 2008).

The CFI's financial resources come to it primarily, or even exclusively, via budgetary transfers from the federal government. But these contributions are meant to be, and are considered to be, endowments. Thus, the full amount is not expected to be expended by the foundation in the fiscal year in which the endowment is given. The government transfers to the CFI and to other foundations (see further discussion below) are deemed, in accountancy terms, to be an 'expense' by government. They are classified as 'conditional grants.' The portion not expected to be distributed in the short term is invested, with the returns on investment constituting additional revenue for the foundation. Once in the bank account of the CFI, these funds are no longer public money under the authority of federal cabinet ministers.

Until recently the Auditor General has not audited CFI spending. The best the AG has been able to do, and it has done so now on three occasions, is to conduct an audit of the government's design of the 'governance frameworks and accountability regimes' of these foundations and to follow up with an audit of the government's subsequent response to the AG recommendations for improvements in these matters (Auditor General 1999, 2002, and 2005). The AG has no mandate to follow the 'money trail' of the public funds given to these foundations in order to conduct a value-for-money or performance audit. Yet, the Auditor General of Canada, Sheila Fraser, has called the management style of this new agency into question, because a $3.6 billion foundation is not subjected to federal Access to Information legislation.

Mandate

The CFI's mandate is to strengthen the capacity of Canadian universities, colleges, research hospitals, and non-profit research institutions to carry out world-class research and technology development that benefits Canadians. The CFI enables institutions to set their own research priorities in response to areas of importance to Canada. According to the Canadian government, this situation 'allows researchers to compete with the best from around the world, and helps to position Canada in the global, knowledge-based economy' (CFI 2008). CFI support is intended to:

- Attract and retain highly skilled research personnel in Canada;
- Stimulate research training of young Canadians;
- Promote networking, collaboration, and multidisciplinarity among researchers;
- Ensure the optimal use of research infrastructure within and among Canadian institutions.

The primary objective of CFI is 'creating the necessary conditions for sustainable long-term economic growth, including the creation of spin-off ventures and the commercialization of discoveries' (CFI 2008). Together with NSERC, SHRCC, and the CIHR examined in chapter 4, CFI must be understood as part of the federal institutional network for R&D.

In the context of R&D, the Government of Canada has stressed the importance of investing in people and also the relevance of post-secondary education for the R&D system. In the CFI's words 'Canada's post-secondary educational institutions and research hospitals play a key role in the innovation process. As both discoverers of knowledge and transmitters of that knowledge through teaching and research apprenticeship, they are an integral part of the "root system" that feeds the country's knowledge base. This root system needs to be nourished' (CFI 2008).

Investing in people includes two ways of attending to the needs of human resources training in Canada. The first is the creation of opportunities for Canadian researchers and students. The ability to attract and retain highly qualified scientists and researchers depends increasingly on the capacity of our post-secondary institutions and research hospitals to provide adequate equipment and research facilities. Leading-edge science and research requires a leading-edge research environment. 'There is a growing concern among Canada's educational and research institutions that we are losing many of our best scientists and researchers to other countries, particularity the United States, because of a lack of adequate opportunities to carry out research in Canada. Investing in research infrastructure – that is, the installations and equipment needed to carry out research – will create the opportunity for Canadians, especially young Canadians, to pursue their careers in Canada' (CFI 2008).

The second way of investing in people is through the development of a technologically oriented workforce. In this case, the concern is not only the loss of researchers to other countries, but also the endowing of students in post-secondary education with the technological capabilities needed to understand the new technologies so that they are able to use them in an effective manner. 'The demands of a technology-driven marketplace require not only engineers, scientists and technicians, but also managers who are technologically knowledgeable' (CFI 2008).

Firms, whether in the health care industry, the 'high tech' sector, or the traditional manufacturing industries, recognize that competitive advantage is based on their ability to innovate and apply the latest 'know-how' to products, processes, and services. 'They [industries] want research-oriented analytical problem-solvers who are a source of new ideas, understand the innovation process and are at ease in the world of new technology. If they cannot find them in Canada, they will tend to take their investments elsewhere. At the same time, firms that aim for future economic success need to be prepared to make long-run investments in skills, education and knowledge – and to provide employment opportunities for young Canadians. The public and private sector share an interest in ensuring that our post-secondary educational system generates the highly skilled workforce the future will require' (CFI 2008).

The CFI has introduced important changes in how public funds are invested in science and technology research infrastructure. As noted above, CFI funds up to 40 per cent of a project's infrastructure costs. The host institution, which is most often a university, raises the remainder. In effect, the CFI's funds are leveraged through partnerships with eligible institutions in the public, private, and voluntary sectors. Levered funding is itself a system of regulation or money with rules and conditions (Foley 1999).

Additional innovations relate to the CFI's grant programs. Traditionally, granting institutions fund individuals or groups of researchers and their research projects. The CFI's programs, however, are directed at *building capacity in the institutions* that conduct research.

The CFI was the first granting body to introduce a requirement for a *strategic research plan* in order to receive CFI funding. Some universities already had such plans but now all universities had to develop them and present their applications to the CFI in the context of such plans. These strategic research plans had to outline the overall research directions of the recipient institutions' research agenda, and also form the basis for reporting on progress. The strategic research plans are the main instrument that the CFI uses to audit funding provided to the universities. Through annual reports, the CFI requests that universities submit information on achievements, progress, and constraints to advance research infrastructure (see below an analysis of three university reports). As we show in the next section, Canadian universities have taken this as an opportunity to adjust to a competitive environment. Some universities in particular have reshaped their goals through this strategic plan process and moved ahead to focus on new areas of research.

As noted above, the CFI funds research infrastructure and not research itself. Research infrastructure includes state-of-the-art equipment, buildings, laboratories, and databases required to conduct research. Eligible institutions can apply for CFI support. Applications are assessed using the following criteria:

- Quality of research and need for infrastructure;
- Contribution to strengthening the capacity for innovation;
- Potential benefits of the research to Canada.

CFI support is awarded following a thorough *merit-based* assessment process (see further discussion in the third section of this chapter) that involves researchers, research administrators, and research users from Canada and abroad who review proposals and make funding recommendations. These volunteers are selected based on their expertise and reputation.

Initial Research Programs

The CFI invested in projects through several initial funding programs:

The *Innovation Fund* enables eligible institutions to strengthen their research infrastructure in priority areas – as identified in their strategic research plan. The fund promotes multidisciplinary and inter-institutional approaches, and enables Canadian researchers to tackle groundbreaking projects. The Innovation Fund helps institutions strengthen their research and training environment in areas of priority that they themselves identify. According to CFI:

> The 2004 Innovation Fund competition was specifically aimed at taking the research capacity of Canadian institutions to the next level by challenging them and their researchers to reach for new heights of excellence, improve their competitiveness, and attain international leadership for the benefit of all Canadians. Institutions were invited to submit proposals that:
>
> - Promote innovative research;
> - Capitalize on excellent research opportunities by drawing on local, national, and international intellectual capital;
> - Help attract and retain top-quality researchers and create a stimulating training environment;
> - Support the development of novel research instrumentation;
> - Create opportunities to fuel the development of clusters and the commercialization of research results;
> - Lead to public policies and improvements in society. (CFI 2008)

The *Infrastructure Operating Fund* was created to assist institutions with the incremental and operating costs associated with the new infrastructure. It contributes to the incremental operating and maintenance costs associated with the infrastructure projects funded by the CFI. Under this funding program, the CFI has invested $135.2 million.

The *Research Hospital Fund* is designed to contribute to research-hospital-based projects that focus on innovative research and training. It supports large-scale infrastructure projects that take a multidisciplinary approach – involving biomedical, clinical, health services, and population health research. According to CFI, 'To address today's and tomorrow's health challenges, support is needed for research that will lead to groundbreaking discoveries and knowledge that can be translated into improved health for Canadians, more effective health services and products, and a better health care system' (CFI 2008).

The *New Opportunities Fund* provides infrastructure support to newly recruited academic staff. The fund helps universities attract high-quality researchers in areas that are essential to the institutions' research objectives. According to CFI, 'The New Opportunities Fund continues to play a crucial role in helping universities to develop and enhance their research training capacity, as well as their ability to transfer knowledge. This Fund provides infrastructure support to new faculty taking up their first full-time position at a Canadian university' (CFI 2008).

The *Canada Research Chairs Infrastructure Fund* provides infrastructure support to the Canada Research Chairs Program. The Program is establishing 2,000 world-class research positions at Canadian universities. Massive faculty renewal – along with growing student demand – was projected to lead to the hiring of as many as 30,000 new researchers over a decade.

In 2000, the CFI established two *International Funds*, each with a one-time $100 million budget. The Canadian portion of projects that qualified under both these funds was eligible to be financed up to 100 per cent. 'In the global race for knowledge, the CFI's International Funds enable Canadian Institutions and their researchers to perform research at the highest international standards of excellence and to partner with some of the best from around the world. The CFI selected nine large-scale research infrastructure projects under its International Joint Ventures Fund and International Access Fund. These projects are aimed at furthering Canada's scientific position in areas such as marine and environmental sciences, infectious diseases, astronomy, light sources and particle physics' (CFI 2008). In 2003–04, the funding for seven of these projects was announced, for a total of $137.5 million.

According to CFI, the *International Joint Ventures Fund* enabled the establishment of high-profile research infrastructure projects in Canada in order to take advantage of unique research opportunities with leading facilities in other countries. On the other hand, the *International Access Fund* provided support to Canadian institutions and researchers to enable them to participate in major international collaborative programs and facilities in other countries.

Finally, the CFI also contributes to human resources training in the context of the knowledge-based economy through awards such as the Career Awards and the Exceptional Opportunities Awards.

Career Awards: In partnership with the Natural Sciences and Engineering Research Council (NSERC) and the Canadian Institutions of Health Research (CIHR), the CFI attempts to recognize the achievements and contributions of Canada's top researchers. According to the CFI, 'Awarded on a competitive basis, the CFI Career Awards acknowledge the important role that state-of-the-art equipment plays in enabling our top researchers to make discoveries in Canada' (CFI 2004,

12). For example, in November 2003, the CFI announced an investment of almost $1 million to provide five NSERC Steacie Fellows with the infrastructure required to carry out their research. The CFI also provided Career Award support to two recipients of the CIHR Distinguished Investigator Award, announced in March 2004. Up to $1 million is available for each of these programs (CFI 2004, 12).

Exceptional Opportunities Awards: The CFI Board of Directors has identified the need for a rapid-response mechanism that will further enable institutions and their partners to participate in unique opportunities for exceptional and innovative infrastructure projects. The first project approved in 2003–04 under this new mechanism was the Structural Genomics Consortium led by the University of Toronto, which was awarded $7.2 million (CFI 2004, 13).

Program Evolution and Evaluation

During 2007, as it approached a program consultation exercise with universities and research institutions with the newly available $510 million from the Harper Conservative government, the CFI reiterated its operating philosophy after a decade of funding. It stressed that the CFI:

- works directly with institutions;
- supports all disciplines in the research and development spectrum;
- uses experts from relevant fields in a rigorous and independent merit-review process;
- requires that applications for funding align with an institution's strategic plans;
- encourages leveraging of federal resources through partner funding;
- has the financial flexibility to negotiate multi-year funding with institutions, to attract funding from other partners. (CFI 2007, 1)

In the view of the CFI, the consultation process 'is ultimately intended to identify opportunities for better alignment with community needs and with the recently released federal S&T strategy' (CFI 2007, 2).

The CFI noted how in 2006 it had implemented its new suite of programs, a program architecture which overall was a response to a series of shifts of emphasis:

- From attraction to *attraction and retention* of top-quality research personnel;

- From investing in new infrastructure projects to *building on and enhancing* the more successful and productive activities enabled by past CFI investments while continuing to invest in *new initiatives*;
- From promoting institutional planning to strengthening *regional and national planning in addition to institutional planning and priority setting*;
- From allocating funds only on the basis of *open competitions to also supporting strategic investments through 'managed' competitions*, including partnership with the federal granting agencies. (CFI 2007, 2; CFI's emphasis)

The titles of the new suite of programs alone were indicative of new perceived needs, but undoubtedly also of new kinds of pressure from universities and from the federal government. As the 2007 consultation process began, the CFI program suite included: the Leading Edge Fund, The New Initiatives Fund, the National Platforms Fund, The Leaders Opportunity Fund, The Infrastructure Operating Fund (IOF), and the Research Hospital Fund (CFI 2007, 4).

Some of the basis for the shift in the CFI suite of programs can also be found in its program evaluation reports, which were a part of its commitment to accountability. Space only allows coverage of one of these reports, namely, an evaluation in 2003 by Bearing Point (formerly KPMG Consulting) of three initial CFI programs, including the previously mentioned Innovation Fund. First, it is useful to note that the evaluation was based on a multidimensional methodology that included CFI progress reports and files, interviews with various advisory and expert committee members that review CFI applications, interviews with university departments heads and/or Deans, and case studies (Bearing Group 2003, 1).

Among the main findings and conclusions reached in the evaluation are:

- The ability of CFI applicant institutions to find the necessary 60 per cent matching funds indicates agreement from external partners (especially the provinces) of the importance of these programs, and most of these funds have been incremental;
- Significant ongoing support will be required. If anything the need will likely increase as applications from the social sciences and humanities rise;
- Overall, the programs have had marked positive impacts ... Where more than half of the infrastructure in the case studies was poor or fair/poor prior to the awards (and none world class), 90 per cent of case study

respondents now rate it as excellent or world-class in the disciplines affected;
- The research that is carried out is more cutting-edge research, conducted faster, with more multidisciplinarity, and with substantially more collaboration (nearly twice as much as before);
- Smaller institutions in particular reported increased visibility and credibility both nationally and internationally as a result;
- It is too soon to measure impacts on research productivity (e.g. through methods such as bibliometrics).
- The majority of projects enabled by the CFI have also increased the ability to attract researchers, postdoctoral fellows and students.
- Both implementing the projects and finding financial resources for operations and maintenance has been problematical in many institutions;
- There are concerns developing in some provinces in terms of the lack of provincial input to research infrastructure planning and decision making;
- Companies that have contributed to purchase costs and/or ongoing research costs expected access to intellectual property or expertise to help in product and process development, access to HQP, and development of more or better relationships with researchers and their institutions. (Bearing Point 2003, 2–4).

Overall, the Bearing Point evaluation concluded that, as of 2003, there were three strategic considerations that CFI faced regarding such programs:

1) Maintaining long-term sustainability will require institutions to convince their provincial partners to supply matching funds, and institutions to find O&M support for the long term. This is the most important long-term strategic issue by far.
2) Additional opportunities for CFI to act as a catalyst for pan-Canadian strategic planning related to research infrastructure should be investigated, possibly including opportunities to act as 'the Canadian voice' in these matters internationally ...
3) CFI and the Social Sciences and Humanities Research Council should continue to investigate ways to encourage involvement in CFI from researchers in the social sciences and the humanities. (Bearing Point 2003, 4–5)

These conclusions were similar in nature to an evaluation of the New Opportunities Fund prepared in 2002 by Hickling Arthurs Low.

This study concluded that there had been positive impacts from CFI funding, but noted as well that if this fund was 'better integrated with support programs from the federal granting councils, it would be easier for Canadian institutions to offer more competitive start-up programs to attract outstanding researchers' (Hickling Arthurs Low 2002, 2).

CFI Funding and the Regulation of Canadian Universities: Three Illustrative Glimpses

To complement the broad account presented above of CFI programs and their initial evaluation, we now look briefly at two small universities, Saint Mary's University and Brock University, and at Canada's largest university, the University of Toronto. Illustrative glimpses of how they see CFI funding and regulation via strategic plans and levered funding can be found in a sample of the annual institutional reports each recipient university is required to provide. In these kinds of brief institutional reports, the universities are invited to comment on how CFI investments are helping in: attracting and retaining high quality faculty, enhancing the training of students and technical personnel, promoting collaborative and multidisciplinary research, promoting cross-sectoral partnerships, and ensuring optimal use of the infrastructure. Universities also have to submit more detailed reports on individual projects, but it is the annual overall institutional reports that provide a broader, albeit extremely brief, glimpse of each university's views.

Small universities like Brock University and Saint Mary's University, whose responses we look at first in this section, reflect the high impact of CFI on their transformation into more focused research universities. For example, Saint Mary's University has moved ahead in terms of defining a strategic research plan and CFI has contributed to its development. In the self-assessment evaluation done, the University stresses, 'Infrastructure support from the Canadian Foundation for Innovation is crucial for Saint Mary's University to achieve the aims of our Strategic Research Plan' (Saint Mary's University 2001, 1).

As noted in its Strategic Research Plan (SRP), Saint Mary's University 'supports high-quality research that builds on existing strengths and partnerships with particular emphasis on interdisciplinary endeavours. We are developing research programs, which are relevant to and engage the Atlantic Canada Community. Our faculty, researchers and students engage in projects that offer value nationally and internationally. It is important to Saint Mary's that faculty, students and the community benefit from research activities at the University' (Saint Mary's University 2005). The priority areas selected by Saint Mary's University to move ahead in the knowledge agenda are Astronomy and Computational Science, Research for Atlantic Canada Communities, and Environmental Studies.

Among the achievements highlighted by Saint Mary's University, the federal Canada Research Chair Program is also cited as significant. For example, thanks to the support of CFI, Saint Mary's University was able 'to recruit two world-class computational astrophysicists to join our Department of Astronomy and Physics' (Saint Mary's University 2004, 1).

Brock University in its own Strategic Research Plan (SRP) identified its transformation from 'a primarily undergraduate, liberal arts university to a comprehensive, more research-intensive university that values excellence in teaching equally with high quality, internationally recognized research and scholarly activity' (Brock University 2004, 1). In order to achieve that goal, Brock University has identified three major objectives: 'facilitating research in all disciplines, with a focus on interdisciplinary research; building on existing strengths and the development of innovative research niches; and knowledge mobilization to the benefit of the University, the Niagara region, and Canada' (Brock University 2004, 1).

In particular, Brock University recognized the value of the CFI's New Opportunities Fund. According to the university, 'this fund has allowed Brock to provide research infrastructure to researchers in the humanities and social sciences as well as the natural and physical sciences' (Brock University 2004, 1). To be sure of the importance of this special fund for Brock University, in the institutional report submitted to CFI in 2004, Brock 'urges the CFI to consider making additions to the New Opportunities pool. Without additional funds to provide infrastructure, it will be increasingly more difficult for regional institutions such as Brock to compete with larger universities to attract and retain highly qualified, new scholars' (Brock University 2004, 1).

In the field of human resources training and recruitment, Brock University has remarked that 'the CFI, OIT and CRC investments have been critical in attracting and retaining HQP, particularly scientists working in the United States and Europe ... [F]or Brock University, the infrastructure has been successful in enabling us to retain our scientists in the face of active recruitment from other Canadian and U.S. institutions' (Brock University 2003, 2).

Not surprisingly, the University of Toronto as a very large and globally high-ranked research university exhibits a considerably different set of responses than the above two small universities in its annual institutional reports to the CFI. First, it notes in its 2005 report that it had by then received grants for 321 CFI-funded research projects amounting to $180 million (University of Toronto 2005, 1), and a year later the total had reached 367 projects. It also made explicit reference in its 2006 report to the fact that the these CFI funds were matched by the Ontario Research Fund and stressed that 'funding from the CFI, matched by the Province of Ontario, has become of fundamental importance in realizing the potential for research achievement in Ontario and Canada' (University of Toronto 2006, 1).

With respect to its strategic research plan, the 2005 report indicates that all four areas of that plan – humanities, social sciences, science and technology, and health and life sciences – were advanced by CFI project funding (University of Toronto, 2005, 1). However, the examples the 2005 report highlights all dealt with research in the last two of these areas of the plan.

Interviews by the authors with some CFI senior staff reinforce the existence of varied impacts on the strategic research plans of large and small Canadian universities. While large and traditional universities already had well-established units of research, small universities in particular seem to have taken great advantage of the support and funding of CFI and its strategic planning requirements to develop and reorient their research priorities.

The above glimpses afforded by the institutional reports also show inherently how, apart from its requirement of a strategic research plan, the regulatory role of CFI, is still limited to the previous goals and objectives as defined entirely by the university. Moreover, the capacity of universities to play the levered money game varies greatly, as does the degree of support from provincial governments. Atlantic Canada universities such as Saint Mary's University have less 'lever' power, in part because the regions' provincial governments have less money per capita than rich industrial Ontario or energy-rich Alberta.

With regard to private sector involvement as partners and co-funders, CFI's 2005–06 annual report indicated that about '38 per cent of researchers reporting to the CFI received funding from industry' (CFI 2006, 11). But private sector funding tends not to be highlighted in CFI reports and evaluations, and thus the picture that emerges indirectly is that the main partners are other governments. In his 2006 presentation to the House of Commons, the CFI's CEO was able to give some specific examples of ways in which the CFI has had positive impacts on the manufacturing sector, but otherwise private sector impacts were more indirect (Phillipson 2006, 1–2. This finding with regard to the basic lack of willingness of the private sector in Canada to invest in R&D was also stressed in chapter 1, and is developed further in the analysis in chapter 6 of universities and commercialization activities.

As noted earlier, the CFI's contribution to the funding of research infrastructure might also put pressure on other funding agencies such as NSERC and the CIHR in particular. The granting council mandate is to provide a constant supply of funding to cover the operating costs and technical experts to keep the infrastructure working. To be sure, an adequate balance between innovative infrastructure and support and maintenance of new equipment is required to assess the impact of CFI's regulatory role vis-à-vis universities (see further discussion below).

The CFI and Federal Research and Innovation Policy

With the CFI mandate and program evolution and evaluation established, and with the reports of the three diverse universities briefly examined, we can now relate the CFI back to the ideas and policy discourse surrounding several issues introduced in chapter 1, namely, the knowledge-based economy, innovation policy, the accountability of foundations as research boundary organizations, the New Public Management, and differences between peer review versus 'merit review' as introduced in chapter 1.

The first powerful influence on the CFI's mandate was, and largely remains, the rapidly rising policy priority given to innovation driven by research and development and underpinning the knowledge-based economy (KBE).

The innovation paradigm is a demanding public policy paradigm wherein priorities such as skills development, education, and research have become increasingly linked to information-era growth and economic competitiveness. Several decades of growth in demand outside universities for university research and for a highly educated work force are seen as evidence of how advanced industrialized nations, including Canada, are transitioning toward a knowledge-based economy (Davenport 2002). A central theme associated with the concept of the KBE and more recent innovation policies is that research and experimental development can 'future proof' national economic and social progress.

Supporting these views are related macroeconomic theories that link national productivity and prosperity to the production and adoption of innovation, through the iterative processes of research and experimental development and through technology-driven economic transformations (Phillips 2007; Lipsey et al. 2005). The ideas and theories of innovation have powerfully embedded new meanings, purposes, and values in Canada's publicly funded national science programs, and in the institutions tasked to implement these policies.

To keep this 'shift' in perspective, it is important to note that in one way or another, and from the earliest times, economic issues have always loomed large in Canadian science and technology policy. The reasons for this is 'Canada's size, its low population density and small proportion of secondary industry,' which led early on to an acceptance that government-supported scientific activities should focus on 'practical applications and the exploitation of the county's natural resources' (Atkinson-Grosjean et al. 2001, 6). While historically Canadian universities have been focused on basic research and education and the training of highly qualified S&T personnel, innovation policies are targeted at the development of new products and processes, particularly in the area of emerging technologies. Metrics such as rates of patenting and the creation of spin-off companies are emphasized. Ideas and theories related to how research is organized and conducted, where it is undertaken, and what it is directed at, are critical to realizing the potential benefits of innovation and the KBE.

More and more often, as the CFI's origins and evolution make clear, public investments in research and development and in building a national innovation capacity are based on systems models of the knowledge creation, dissemination, translation, and the commercialization process. Concepts and ideas related to national and local systems of innovation, and clusters as complex multi-directional interactions among firms, universities, and government and their variously networked S&T personnel, form the overall discourse influencing policy and funding action (Doern and Kinder 2007; Edquist 2004; Boden et al. 2004; Freeman and Soete 1997).

It is argued that business and jobs will 'cluster' around talent (Courchene 2004). This talent, which is often found in universities, will create the innovative products and processes that, when adopted (and where possible commercialized), will lead to improved social conditions and to economic prosperity.

The uptake of these ideas has contributed to new mandates and roles being assigned to the increasing number of federal institutions that support research and skills development. In this regard, as chapter 4 has already shown, a shift in the mission and roles of the research granting councils (NSERC, SSHRC, and the CIHR) was justified by the need to make more targeted investments in human capital and research which, of course, is largely based in universities. These ideas and practices are also expressed as a shifting away from the former principles and values of bottom-up, investigator-driven emergent discovery towards greater emphasis on the funding of strategically driven, multisectoral, and multidisciplinary alliances of applied research.

Universities are being asked to accept new partnerships with industry and government, but as industry and government are oriented towards different goals this might create tensions with universities. Certainly the emergence of the KBE has added a new challenge for Canadian universities in a context of global competitiveness. In liberal democracies there has been a shift over the last century that can be accurately characterized by significant degrees of separation and segregation between the university, the state, and the market. Recently, however, as chapter 1 has shown, it has been posited that the balance is shifting away from relative autonomy towards a new 'mode of knowledge production' (Gibbons et al. 1994), of which the growing engagement of universities with their regions and localities, along with the greater use of interdisciplinary research, is an important aspect.

The CFI, as the first Canadian innovation-era boundary foundation, was 'the direct result of lobbying by a handful of university presidents – lobbying directly at the Prime Minister's Office with the support of Kevin Lynch, then Deputy Minister of Industry Canada [and subsequently Deputy Minister of Finance]' (Doern and Levesque 2002, 136–7). As chapter 3 has shown, early lobbying by the AUCC also played a key role.

The Department of Finance was the birthplace of these new foundations for two reasons. First, there was the demand for substantial new funding; second, there was the demand to protect funding from other cuts. The expenditure budget systems in place when budget surpluses first emerged in 1997 left new spending entirely in the hands of the PM and his Finance Minister. Second, it was Finance that would have to devise a mechanism to lock away money in a way that would meet the 'generally accepted accounting principles' (GAAP) set by the Public Sector Accounting Board of the Canadian Institute of Chartered

Accountants (the premier professional body of public accountants in Canada). The method chosen was to give the money as endowments to organizations not under the control of government. This locked the money away. It also conveniently allowed the government to avoid putting the entire budgetary surplus into debt reduction. Not surprisingly, as noted earlier, the Auditor General has strongly criticized this method of financing public policy programs (Canada 2001, 2005; Auditor General 2002, 2005).

According to Aucoin, by adopting this independent foundation model with the requirement that the foundations be outside the control of government, those responsible were also able to secure another set of objectives: to shield the foundations from both the performance audit mandate of the Auditor General and the statutory access to information regime (Aucoin 2003). Thus, the core accountability values discussed in the core values and ideas part of chapter 1's framework were at stake. For example, the senior officials in the Departments of Industry and Finance, especially in designing the prototype agency, the CFI, wanted an organizational mechanism that was even more arm's length from the government than the various non-departmental bodies that were already part of the Industry Canada stable of agencies (Doern and Levesque 2002, 136–7).

We argue that the creation of autonomous boundary organizations like the CFI is indeed crucial, so that funding allocations are not 'political' in the sense of being determined by partisan political criteria per se. Such funding does require arm's-length assessment and advice by experts. There is, however, a difference between peer review as required by the granting councils and 'merit-based' review as practised by the CFI. Scientific, technical, and research peers are involved in CFI decisions about funding projects, but merit review is considerably broader in that the merit-based process involves assessment and advice by other kinds of non-science expertise from university and institutional managers regarding broader aspects of infrastructure construction, management, and varied notions about what research 'capacity' might mean in very particular project situations and configurations.

We have also had glimpses in earlier sections of this chapter that the boundary between what is 'infrastructure' and what is 'operational' and how much is funded is an issue not only at the project level but also in the changing program architecture or program suite. As we have seen, special operational funding add-on programs were added to the CFI's repertoire in much the same way that federal overhead funding for universities regarding granting council grants is a recent add-on. Not only are such boundary concepts inherently inexact, but the CFI itself as patron may well have an initial bias in defining things narrowly so that more project grants are allocated in ways intended to highlight the program's positive impacts nationally and regionally. In a sense, there remains a considerable messiness and arbitrariness to Canadian policy regarding the overlapping issues of the indirect, overhead, and capital costs of research. In the final analysis these are often governed, as in other policy-spending fields, by the rough calculus of how to parcel out funds among the needy and the deserving.

In this regard, the CFI's institutional configuration attempts to deal with the patron and regulatory role of the federal government in R&D. In reality, CFI is located on the border between politics and science, where its mission is defined as a provider of infrastructure funding. As chapter 1 has indicated, the creation of independent agencies in the field of science policy has been analysed, by authors such as Guston, in terms of the creation of boundary organizations (Guston 2000). Such boundary organizations, insulated from political authorities, relate to the need for politicians and researchers to provide an accountability framework for R&D activities and to resolve problems of asymmetrical information between patrons and researchers. Based on technical expertise, these boundary organizations 'engage in such activities to exploit opportunities or respond to threats from their environment. The boundary organization ... draws its stability not from isolating itself from external political authority, but precisely by making itself accountable and responsive to opposing, external authorities' (Guston 2000, 32). It is in this sense that agencies like CFI occupy the spatial border between science and politics.

As an independent foundation, CFI also has all the features specified in the concept of the New Public Management introduced in chapter 1. According to Aucoin, it is part of the process of the 'internationalization of agencification' as a NPM innovation in governance among OECD countries in that it has three main characteristics: autonomization, since it does not belong to any specific ministerial branch; 'contractualization,' since their staff work under contracts and they belong to the permanent body of public agents; and finally; 'dissaggregation,' in the sense that they are highly specialized units in particular areas of policy (see Aucoin, 2005). It has also sought to ensure more direct kinds of accountability downwards and outwards to its institutional clientele as well as upwards to the federal government via its funding agreement and other reporting to the House of Commons.

The Canadian Foundation for Innovation, the Canadian Millennium Scholarship Foundation, and the Foundation for Sustainable Development Technology were established by legislation. Over 70 per cent of all federal funding to foundations goes through these three organizations. Other foundations, such as Genome Canada, were created under the Canadian Corporations Act. Some foundations must maintain their endowment[2] and only invest the revenue the endowment generates (for example, the Pierre Elliott Trudeau Foundation). Others, such as the Canadian Foundation for Innovation, can use both the principal and the investment revenue. Today in Canada there are fifteen independent foundations.[3]

Conclusions

This chapter has examined the Canada Foundation for Innovation as both a patron and a regulator, and the ways in which the foundation has changed the relationships between the federal government and universities. Although the CFI is crucially an arm's-length foundation and is independent of politics in the sense of partisan politics, the analysis has shown that it does have to deal, as all patrons do, with small-p politics – among universities in the form of regional distributional politics, among different realms or disciplines of science and their varying infrastructure needs, and among small and large universities.

In addition, we have argued that the CFI acts as a *regulator* because investments in research and development infrastructure must conform to certain 'rules' set by the CFI, including requirements that universities submit strategic research plans and also that funding be levered. In short, universities do not get infrastructure money unless they bring money in the form of funding commitments from other funders, public

---

2 'Endowment' is the term used for monies that are transferred from the federal government to foundations.
3 Canada Foundation for Innovation, Canada Millennium Scholarship Foundation, Canada Health Infoway, Genome Canada, Aboriginal Healing Foundation, Green Municipal Investment Fund, Canadian Health Services Research Foundation, Pierre Elliott Trudeau Foundation, Canada Foundation for Sustainable Development Technology, Canadian Foundation for Climate and Atmospheric Sciences, Clayoquot Biosphere Trust Society, Forum of Federations, Pacific Salmon Endowment Fund Society, Canadian Institute for Research on Linguistic Minorities, and Frontier College Learning Foundation.

or private. The chapter has suggested that universities have widely differing capacities to raise or obtain these levered funds.

Aspects of regulation are also evident in the realm of defining what is eligible for funding, and hence in defining what in fact 'infrastructure' is as capital rather than operational spending. Such issues inherently arise in the CFI's role of developing research 'capacity,' and in determining what 'innovation' means in these funding choices and what kinds of research capacity-building might occur when CFI funds are allocated.

Some of the 'transformative' role of the CFI can be seen in the recruitment/retention of faculty, in the enhancement of universities' research experiences through the submission of strategic plans, and, perhaps more importantly, in the promotion of research infrastructure per se. The CFI introduced two important shifts in the context of the Canadian R&D system overall. The first was the introduction of a required 'strategic research plan,' which implies the definition of projects in the context of economic and related research priorities. The second was moving away from funding individual researchers to funding institutions such as universities and larger interdisciplinary teams of research partners.

The analysis has shown that the impact of the CFI in Canadian universities seems to be varied. On the one hand, large universities have garnered advantages from the new funding provided, but already had a long-standing tradition of research. On the other hand, in order to become more competitive, smaller universities with a less-developed tradition in research and development have taken this CFI support as an opportunity to restructure research capacities and rebuild their academic profiles.

With regard to other overall links to federal policy, we have argued that the CFI's creation was the result of two related trends that started to develop in the 1980s and more systematically in the 1990s within the Canadian Government. On the one hand, it was a consequence of the appearance of the KBE as a core concept tied to the achieving of global economic competitiveness and linked closely to innovation as a policy paradigm. On the other hand, it was due as well to the adoption of ideas about boundary organizations, and to the New Public Management.

As a boundary organization, CFI can be seen as a partial 'solution' to the problem of asymmetrical information between politicians and researchers. By providing autonomy and technical expertise to independent agencies, politicians can guarantee a sustainable transfer of funding to the R&D system, in this case on matters of research infrastructure. As an NPM 'agency' the CFI has sought to foster closer evaluated links between itself and universities, as well as with other research agencies such as hospitals. In terms of infrastructure decisions and assessments of capacity, this has meant a broadened reliance on merit-based assessment rather than on peer review only.

The university system overall became, via the CFI (and other federal research and innovation policy changes), more prominent in the new economic strategy pursued by the Federal Government. Universities were pressed to accept and promote new partnerships – 'a new social contract' – among universities, industry, government, and even the voluntary sector on infrastructure issues and overall strategic research plans.

As we have seen, the management and accountability of the CFI as an independent agency has been criticized especially in its earliest years. However, the CFI can argue with some validity that its activities are subjected to many different oversight mechanisms, including the requirement of the tabling of annual reports through the Minister of Industry. In addition, reviews, audits, and other types of reports are subjected to scrutiny from parliamentary committees.

REFERENCES

Atkinson-Grosjean, Janet, D. House, and D. Fisher. (2001). 'Canadian Science Policy and Public Research Organizations in the 20th Century.' *Science Studies: An Interdisciplinary Journal for Science and Technology Studies* 14 (1): 3–25.
Aucoin, Peter. (2003). 'Independent Foundations, Public Money and Public Accountability: Whither Ministerial Responsibility as Democratic Governance?' *Canadian Public Administration* 46 (1) (Spring): 1–26.
– (2005). 'Accountability and Coordination with Independent Foundations: A Canadian Case of Autonomization of the State.' Paper presented to Workshop 'Autonomization of the State,' Scancor and the Structure and Organization of Government Research Committee, International Political Science Association, Stanford University, Stanford, California, 1 and 2 April 2005.
Auditor General. (1999). *Report to the House of Commons*. Chapter 23, 'Involving Others in Governing: Accountability at Risk.' Ottawa: Public Works and Government Services Canada.
– (2002). *Report to the House of Commons*. Chapter 1, 'Placing the Public's Money beyond Parliament's Reach.' Ottawa: Public Works and Government Services Canada.

- (2005). *Report to the House of Commons*. Chapter 4, 'Accountability of Foundations.' Ottawa: Public Works and Government Services Canada.
Bearing Point. (2003). *Evaluation of the CFI Innovation Fund, University Research Development Fund, and College Research Development Fund*. Bearing Point (formerly KPMG Consulting). May.
Boden, Rebecca, Debora Cox, Maria Nedeva, and Katherine Barker. (2004). *Scrutinizing Science: The Changing UK Government of Science*. London: Palgrave Macmillan.
Brock University. (2001). *CFI Institutional Report* (CFI).
- (2002). *CFI Institutional Report* (CFI).
- (2003). *CFI Institutional Report* (CFI).
- (2004). *CFI Institutional Report* (CFI).
- (2005). *CFI Institutional Report* (CFI).
Canada. (2001). House of Commons Standing Committee on Industry, Science and Technology, *Hansard* 37th Parliament, 1st Session, 3 April. Ottawa: Public Works and Government services Canada.
- (2005). 'Accountability of Foundations: Backgrounder.' Ottawa: Department of Finance. Available from: http://www.fin.gc.ca/toc/2005/accfound-eng.asp (accessed 11 January 2009).
CFI. (1997). *Budget 1997: Building the Future for Canadians*. Canada Foundation for Innovation, 18 February 1997.
- (2004). *2003–2004 Annual Report*. Canada Foundation for Innovation. Available from: http://www.innovation.ca/docs/annualreport/annual04_e.pdf (accessed 11 January 2009).
- (2006). *Solid Foundations: Annual Report, 2005–2006*. Canada Foundation for Innovation.
- (2007). *Discussion Paper: CFI Program Consultation*. Canada Foundation for Innovation.
- (2008). 'Site Map.' Canada Foundation for Innovation. Available from: http://www.innovation.ca/sitemap/index.cfm (accessed 15 December 2008).
Courchene, Thomas J. (2004). 'Social Policy and the Knowledge Economy: New Century, New Paradigm.' *Policy Options* 25 (7) (August): 30–6.
Davenport, Paul. (2002). 'Universities and the Knowledge Economy.' In *Renovating the Ivory Tower*, ed. David Laidler, 39–59. Toronto: C.D. Howe Institute.
Doern, G. Bruce, and Richard Levesque. (2002). *The National Research Council in the Innovation Era*. Toronto: University of Toronto Press.
Doern, G. Bruce, and Jeff Kinder. (2007). *Strategic Science in the Public Interest: Canada's Government Laboratories and Science-Based Agencies*. Toronto: University of Toronto Press.

Edquist, Charles. (2004). 'Reflections on the Systems of Innovation Approach.' *Science and Public Policy* 31 (6): 485–90.
Foley, Paul. (1999). 'Competition as Public Policy: A Review of Challenge Funding.' *Public Administration* 77 (4): 809–36.
Freeman, Chris, and Luc Soete. (1997). *The Economics of Industrial Innovation.* London: Pinter.
Guston, David. (2000). *Between Politics and Science: Assuring the Integrity and Productivity of Research.* Cambridge: Cambridge University Press.
Gibbons, Michael, C. Limoges, H. Nowotny, T. Trow, P. Scott, and S. Schwartzman. (1994). *The New Production of Knowledge: The Dynamics of Science and Research in Contemporary Societies.* London: Sage Publications.
Hickling Arthurs Low. (2002). *Evaluation of the New Opportunities Fund.* Hickling Arthurs Low.
Lipsey, Richard, Kenneth Carlaw, and Clifford Bekar. (2005). *Economic Transformations: General Purpose Transformations and Long Term Economic Growth.* New York: Oxford University Press USA.
Phillips, Peter. (2007). *Governing Transformative Technological Innovation: Who's In Charge?* Northampton, Ma.: Elgar.
Phillipson, Eliot A. (2006). 'Notes for presentation to the House of Commons Standing Committee on Industry, Science and Technology.' 9 November. Canada Foundation for Innovation.
Saint Mary's University. (2001). *CFI Institutional Report* (CFI).
– (2002). *CFI Institutional Report* (CFI).
– (2003). *CFI Institutional Report* (CFI).
– (2004). *CFI Institutional Report* (CFI).
– (2005). *CFI Institutional Report* (CFI).
University of Toronto. (2005). *CFI Institutional Report* (CFI).
– (2006). *CFI Institutional Report* (CFI).

6 Universities, Commercialization, and the Entrepreneurial Process: Barriers to Innovation

PAUL J. MADGETT AND CHRISTOPHER STONEY

The federal government has committed substantial resources in an attempt to steer Canadian universities towards increased commercialization, innovation, and entrepreneurship. It has also sought to exhort universities to move in this direction through studies, reports, and ministerial speeches. This more overt commercialization strategy has been prompted by a changing international context that has seen industrialized countries being challenged by low wages, government subsidies, and market potentials of many less developed, particularly Asian, nations. As noted in chapter 1, the Harper Conservative government in Ottawa has released its new vision for maintaining and enhancing Canada's competitive advantage (Industry Canada 2007). The strategy document examines the entrepreneurial, knowledge, and people advantage needed to create a sustainable competitive economy.

Addressing broadly the same concerns, this chapter focuses on the issues involved and the mechanisms that could be used to further improve the role of university research in commercialization. Continuing the discussion in chapter 1, it further establishes the larger commercialization policy setting for the chapters which immediately follow, namely those on intellectual property and technology transfer offices, university–government collaboration in research, and the co-location of public science on university campuses.

Significantly, Brzustowski conceptualizes a successful entrepreneurial process as requiring both an invention system focusing on intellectual property and a system for commercialization. He also maintains that the success of the commercialization mechanism is dependent upon the ability and expertise of those individuals involved in the technology transfer offices at Canada's universities (Brzustowski 2006a), a key focus of chapter 7 in this volume.

The federal government may further influence this process by expanding the development of public–private research and commercialization partnerships (Industry Canada 2007). This strategy could complement the involvement of small and medium-sized firms in the commercialization process by following the recommendations of the publication 'People and Excellence: The Heart of Successful Commercialization' and creating a Canadian SME Partnerships Initiative (Industry Canada 2006). The need for effective partnerships is also illustrated by Brzustowski's finding that: 'Compared with the most innovative companies in the U.S. and Europe, the best of the Canadian private sector spend a significantly lower proportion of their sales revenue on the R&D to produce new products, and therefore innovate less frequently' (Brzustowski 2006b, 6).

In an effort to regain and maintain a competitive edge, the different levels of government in Canada have recently emphasized the need to develop coherent strategies that utilize the vast research and technological capacity of our universities in the race to provide commercially viable technologies as well as innovations that can be employed to improve industrial processes. In so doing, they follow others who have argued that the country needs to focus on commercialization – not only on improving or creating new processes or products but also on the entrepreneurial process, which 'involves all the functions, activities, and actions associated with the perceiving of opportunities and the creation of organizations to pursue them' (Bygrave and Hofer 1991, 14).

In particular, it is anticipated that an improved entrepreneurial process in universities would help all three levels of Canadian government stimulate innovation by creating new industry groupings in specific areas of science and technology. The long-term objective is to harness this potential for clusters similar to Route 128, Washington, and Silicon Valley. David Wolfe's analysis in chapter 11 focuses on clusters, but it is important to stress from the outset that they are also a part of the larger context for commercialization. The development of geographical clusters is intended to produce a number of important benefits and outputs such as greater research diffusion, stronger technology infrastructure, and substantial spillovers in R&D. These groupings are expected to result in an agglomeration of firms to create higher-paying jobs, which are closely related to increased productivity and competitiveness (Laidler 2002). However, as of yet there is little evidence of cooperation between the different levels of government and this aim still appears utopian, with enduring rifts limiting the potential to better facilitate this type of cluster strategy.

In recent years, for example, federal research granting councils have substantially increased new funding, yet these have not resulted in a larger array of revenue-generating research applications. There are other factors contributing to the current lack of effectiveness, including a dearth of entrepreneurial characteristics, gaps in national intellectual property legislation, too much reliance on public funds as a stable source of revenue for universities, and jurisdictional issues between provincial and federal governments. In addition, we concur with Brzustowski's view that the private sector has not been integrated sufficiently and that more effort should be concentrated on bringing university research to solving industrial problems (Brzustowski 2006a).

In order to address the issue of improving the policy aims associated with commercialization, a number of policy options are also examined, including the establishment of university business and technology parks, the replacement of licensing with equity stakes, the restructuring of research funding, and the outlining of a different set of responsibilities for universities. While this book as a whole focuses on universities per se, some reference to community colleges is also necessary to deal properly with commercialization and innovation challenges. The aim of such changes would be to reduce the barriers to knowledge diffusion, and to increase the coordination and synergies between the two main tertiary education players in order to better enable commercial opportunities and returns on taxpayer investments.

By exploring these opportunities for increased commercial outputs from higher education the chapter also illustrates the potential importance of higher education research and development for the Canadian economy, both as a catalyst for innovation and as a major player in terms of the total research performed. In order to explore the extent and effects of commercialization on the Canadian university sector, the chapter draws on data obtained from Statistics Canada Higher Education Commercialization Surveys conducted since 1998, as well as their estimates for national research and development expenditures from 1993 to 2005. In particular, this data provides a picture of past trends relating to technology transfer and may also identify some of the barriers preventing businesses, universities, and community colleges from developing marketable discoveries.

To this end, the experience of several countries will be referred to briefly in order to contextualize the Canadian scene and to explore alternative policy options. The chapter looks first at the basic context of the commercialization policy issue regarding universities. This is then followed by a section which focuses on three necessarily related questions:

- What are the economic benefits of higher education R&D?
- What is the current state of the key factors affecting commercialization outputs? and
- How can the diffusion of university research be improved?

Context and Background

As chapter 1 has clearly indicated, greater emphasis has been put in the last decade on innovation as the potential bi-product of the commercialization of university research. In defining innovation as 'the process of bringing new goods and services to market,' the federal Advisory Council on Science and Technology (1999, 1) was articulating a growing and often implied belief that in order for knowledge to be of benefit to society, innovation has to be transferred to the market. To this end, research outputs should not be limited to licensing but also need to be re-conceptualized to include various facets of technology transfer and the entrepreneurial process, which would help to increase the number of small and medium-sized enterprises.

This broader view of commercialization per se typically involves two types of start-ups: the first operating in product markets by developing products and services, and the second operating in markets for technology by transferring this knowledge to other firms in order to to develop and sell products and services (Pries and Guild 2007). Consequently, any medium that would allow these products and services to be marketed to the international and national marketplace through some sort of knowledge transfer mechanism would help sustain Canada's technological prowess. Furthermore, for these outcomes to be successful, the government needs policies that will nurture geographical locations that are creative, dynamic, and attractive enough to stimulate entrepreneurial excellence (Youl Lee et al. 2004). According to Fisher and Atkinson-Grosjean (2002), innovation policies usually promote the commodification of knowledge, accelerated development of information-based technologies, and the flow of capital and skilled labour within globalized markets.

This trend of integrating market principles into higher education (universities and community colleges) has been enabled by increased global competition and demand for accountability by governments keen to justify their investments of taxpayers' monies (Newman et al., 2004). In addition, this trend is being accelerated by public funding being diverted to other public programs relating to health care, K–12 education, and crime prevention, thereby limiting the available funds to higher education (Kerr 2001).

Slaughter and Leslie (1997) determined that faculty reacted to the lack of resources by increasing their academic capitalism in order to access further research funds. Similar concerns about the efficiency of publicly funded higher education are not unique to Canada and have been strongly debated in the United States. In fact, according to Noftsinger, our southern neighbours have been carrying on this discussion since the 1980s. It has been associated with conservative reforms that are a by-product of attempts to re-energize basic education and a public belief that universities have become too isolated and far too costly (Noftsinger 2002).

Meanwhile, recent work in the United States has examined citizen attitudes towards university research. As Newman et al. argue, 'the public also values the research that universities perform, most often associating the benefits with medical research. The value of research to economic development is generally recognized, but it is still a vague recognition' (Newman et al. 2004, 71). With analogous Canadian views in mind, the federal government has linked commercialization with the future prosperity of Canada and ratcheted up expectations for output and outcomes from their research investments in the higher education sector. Currently, commercialization has various output indicators: invention disclosures, patent applications filed, patents awarded, licences, revenues, and spin-offs, but some believe that the process is not linear and there is a four to six year lag time from basic research to some substantive output (Langford et al. 2006).

A further limitation on the results of increased government research spending arises from a reluctance on the part of academics to sacrifice their traditional autonomy, independence, and academic freedom. Davenport, a strong advocate of higher education, believes that a balance must be sought between these two mandates: 'Universities are special places because learning takes place in an environment of research and scholarly innovation' (2002, 45). Consequently, even advocates of increased commercialization do not wish to see all resources spent on the current needs of the marketplace, preferring that a sustainable equilibrium be established between research and development, and between pure and applied research. From this perspective, the continued relevancy of higher education research is dependent upon its ability to adjust to the environment it now faces:

> The choice is between holding fast to the past, which promises continued confrontation and ends in decline, and grasping the challenge of change, with all its risks and uncertainties but with the promise of reshaping universities to better suit the needs of our time. (Cameron 2002, 146)

The extent to which universities change in response to the increased emphasis on innovation and commercialization will depend largely on the substantiation and assessment of the benefits.

Three Key Questions

What Are the Economic Benefits of R&D?

Knowledge networks, regions, and/or partnerships are seen to play an essential role in the development of innovation as well as in knowledge diffusion. All post-secondary institutions in Canada provide an economic impetus through their employees and goods and services, which maintain a stable flow of capital into their locality. Also, knowledge agglomeration in the geographical vicinity of universities helps to garner attention from government and the private sector and may create a competitive advantage for those localities.

For many years, governments have identified the knowledge economy as an advantage for industrial countries, especially in their efforts to compete with developing nations. Economic development and higher education have been credited with the ability to unite the various stakeholders of different levels of government, education, business, and industry, enabling them to work aggressively to compete nationally and internationally (Nofsinger 2002). Currently, the manufacturing bases in some Western nations do not have the advantage of cheap inputs such as the land or resources or labour available in some developing countries (Audretsch 2003). Considering the options available to companies to compensate for these constraints, Audretsch listed the following three alternatives: reduce operational costs of inputs, incorporate technology and equipment to maximize productivity, or shift production to low-cost countries. He also noted that these changes were already occurring with the downsizing of companies in Europe and the United States. Canada has been to some extent immune to many of these hardships because of its strong ties with the U.S. economy and, for many years, a depressed currency. However, considerable exchange rate volatility in recent years has caused the productivity advantages and cost savings for many of our exports to erode.

Various internal and external pressures have prompted Canada to explore the benefits of further developing a knowledge-based economy aimed at stimulating innovation. In their geographical innovation research, Audretsch and Feldman have identified certain occurrences such as stronger levels of innovation in California and Massachusetts compared to the weaker outputs found in Midwest. The Midwest in the United States has some prestigious research-intensive universities, such as the University of Wisconsin, the University of Illinois, the University of Michigan, Northwestern University, and the University of Chicago. After controlling for the concentration of production, the authors concluded that university research was a positive and significant factor stimulating geographical agglomeration (Audrestch and Feldman 1996). Although this is difficult to quantify, the entrepreneurial culture of these universities is attributed with a leading role in helping to give Massachusetts and California a competitive advantage over other regions. More specifically, Feldman (1999) has illustrated that the success of innovation in certain geographical locations is intertwined with the types of activity being pursued and the life cycle of the companies pursuing them. In other words, while entrepreneurial culture is not a guarantee of innovation and commercial success, research suggests that it is an essential ingredient.

In many countries, research has underlined a need to concentrate funding on Small and Medium-sized Enterprises (SMEs) to help strengthen local economies. However, these enterprises will, for the most part, tend to adopt already tested technologies due to the limited internal capital available to fund their own research and development (Tödtling and Kaufmann, 2002). Nevertheless, many of the returns remain in the locality and these SME-focused numbers provide an attractive political headline. Tödtling and Kaufmann (2002) also explained that large companies are prepared to absorb greater risks because of the greater long-term benefits.

In order to stimulate these benefits, the role of universities is twofold: the first is to supply the economy with highly skilled personnel, while the second is to provide new knowledge for the economy (Guthrie 2006). Expanding on this concept, there exist two different types of knowledge created in the economy and both are affected by higher education: the first is frontier knowledge mostly created by R&D, while the second is knowledge created from experience and education (Loyd-Ellis 2000). Accordingly, universities may play a more significant role if they move to develop a flexible curriculum by mixing practical experience with educational research and knowledge. That said, Guthrie believes that such knowledge must also be translated into tangible outputs if it is to help Canada confront the challenges of increased competition:

> In an increasingly competitive global economy, we must become better at turning our world-class knowledge into marketable goods and services. We must be more entrepreneurial and responsive to customer needs, whether they originate down the street or on the other side of the ocean. (Guthrie 2006, 3)

In Canada, universities and a grouping of community colleges are playing a significant research role. The percentage of research performed in Canada by higher education grew from 27.5 per cent in 1994 to almost 38 per cent in 2006.[1] In terms of innovation, research suggests this should have a positive impact. Park (1994), for example, concluded that government R&DD (research and development and demonstration) has some effect on increasing adoption by reducing the unknown cost barriers associated with innovation, resulting in a greater level of productivity and innovation in the economy.

Industrial research growth during the same period has slowed from 56.7 per cent in 1994 to about 53 per cent in 2006. The government research funding increase that favoured universities reflected the Liberal government's strong emphasis on increasing national innovation via the university sector. Until recently, there has been no federal funding of indirect research costs, resulting in an increased financial burden for Canadian universities, which, as highlighted in Tupper's analysis in chapter 2, have already suffered from a reduction of provincial operational funding in many parts of the country. In contrast, in the United States, the government provides substantial funding, normally at least an additional 50 per cent, for indirect costs associated with research grants. Some of the provinces, Ontario for example, have begun to fund the overhead costs associated with some of their research programs, such as Ontario Centres of Excellence, but by 2004 only an estimated

---

1 Statistics Canada, Research and Development Expenditures (Statistics Canada 2008).

5 per cent of the reported sponsored research had provisions to cover indirect costs assumed by the financier (Statistics Canada 2004).

What Is the Current State of the Key Factors Affecting Commercialization Outputs?

A plethora of government policies and funding sources have stimulated debate on how to generate greater return for these public funds. This dialogue has been ongoing at different levels of government to provide strategies to help in the diffusion of research through collaborations, partnerships, and other mediums. Roberts and Lloyd-Ellis's model further elaborates on this issue:

> One could re-interpret our R&D sector as an initial adoption sector which procures technologies internationally and engages in their costly adoption. Once the initial adoption costs have been incurred, it becomes possible for the knowledge to disseminate throughout the economy, and eventually to find its way into the education system. (2000, 32)

This discussion has initiated the release of various reports from government departments and policy think tanks such as Industry Canada and the Conference Board of Canada. These reports and other studies suggest that the large investments in research funds and large research-intensive universities are not the sole determinants of success in improving the dissemination of research. Some of the other variables identified include the availability of high-risk funds, the development of a risk-taking culture, and developing strong information and business creation networks (Audretsch 2003).

International research suggests that there are various types of institutional factors that influence the probability of being successful in commercialization and technology transfer. Findings from Italy, for example, have illustrated the need for faculty to become more aware of the external environment in order to explore actively the potential applicability of discoveries (Grandi and Grimaldi 2005). In the United States, a study by Siegel et al. (2003) accentuated the willingness of faculty to disclose their inventions as a key factor increasing the pipeline of discoveries, especially when related to promotion and tenure criteria and university policies on dispersion, royalties, and equity. The Japanese, under former Prime Minister Junichiro Koizumi's leadership, have seen the benefits of commercializing university research and have implemented new policies. These changes include recently giving universities the responsibility of owning and managing their intellectual property rights and quantifying outcomes with new ventures and job growth (Fitagawa 2005).

Even with federal support of both a funding and an exhortative nature, this task may not easily be accepted or acted upon in Canada because of strong resistance by academia and the division of government jurisdiction in post-secondary education. This reticence has been particularly visible in the arts and humanities, resulting in scepticism towards things commercial, perhaps partly because the faculty in these departments can rarely partake in their benefits. Not surprisingly, Hum (2000) found that 78 per cent of spin-off companies created were from science disciplines – biotechnology, health sciences, engineering and applied sciences, and mathematical and physical sciences. Similar studies find that the applied focus of biological sciences and engineering provide an even greater opportunity for technology transfer than traditional physical sciences (Thursby and Kemp 2002).

Bercovitz et al. (2001) studied Duke University, Pennsylvania State University, and Johns Hopkins University, and established that each organization's specific traits affected their institutional ability to transfer technology. In particular, certain latent inefficiencies, associated with their mission of basic research and teaching, appear to hinder a university's ability to fully exploit commercialization (Thursby and Kemp 2002). However, as these roles constitute the traditional *raison d'etre* of universities they are unlikely to disappear even if that places limits on their commercial potential.

A study carried out by Powers and McDougall (2005) in the United States identifies four variables as significant predictors of technology transfer (start-ups formed from intellectual property licenses): industry R&D revenues, faculty quality, age of technology transfer offices, and access to venture capital. Chapter 1 has shown through reference to data in the Appendix that since the mid-1990s Canadian research funds have grown substantially in all three sectors, with the federal and provincial governments playing dominant roles. It also shows Canada's relatively low level of business R&D, and so it could plausibly be argued that the lack of growth in the private sector is partially responsible for Canada's rather mediocre commercialization results. That said, without the ability to isolate the effects of private sector research it is impossible to test Powers's (2004) hypothesis that industry research plays an enabling role in Canada. Moreover, without more detailed data, it is not possible to identify the effects of each sector's funding from its counterparts; consequently, it is difficult at this stage to ascertain what percentage of the overall growth in licences and revenues is caused by the growing sum of monies invested in higher education research.

In order to refine its evaluation of spending on higher education, the Canadian government has begun to collect extensive data on commercialization activity, beginning in 1998 through its new research instrument, the 'Survey of Intellectual Property Commercialization in the Higher Education Sector.' However, this survey, conducted annually since 2003, has changed in recent years; consequently, it is impossible to compare all the data reports in their entirety. For example, as of 2003 university and hospital commercialization indicators have been combined; therefore, in an effort to maintain some uniformity, we are using data generated with the CANSIM access tool as part of the broader government survey referred to above.[2]

Figure 6.1 demonstrates that since 2001 the revenue streams from intellectual property have stagnated. Nevertheless, the number of invention disclosures increased from 661 in 1998 to a 1,353 (preliminary) in 2004, while the number of research contracts increased from 5,081 in 1998 to 11,432 in 2003. The number of patents issued increased drastically from 143 in 1998 to 396 (preliminary) in 2004, while the number of new licences and options increased from 243 in 1998 to 491 (preliminary) in 2004. It can also be noted that in 2003 higher education was making approximately $19 million from its intellectual property (i.e., revenue–operational expenditures). Figure 7 of the Appendix shows as well that in 2005 commercialization outputs were overwhelmingly focused on large universities rather than on small and medium-sized ones.

In order to present further measurements indicating the effectiveness of commercialization in Canada, we draw on a study by Clayman and Holbrook. The authors conducted an analysis of the current status of spin-off companies incorporated in 1995, using a select group of ten medical/doctoral and two comprehensive institutions. This sample was found to have aided in the creation of 301 firms of which 219 were

---

2 Unfortunately, due to uncertainties about the definition pertaining to a few variables, earlier data from a 1997 feasibility study will not be included. Figure 6.1 contains data that includes public hospitals as well as their affiliated universities except for the years 1999 and 2001, when hospital information was confidential. Also, the complete data from 2004 had not been released at the time of writing of this chapter.

Figure 6.1: Commercialization Revenue and Costs

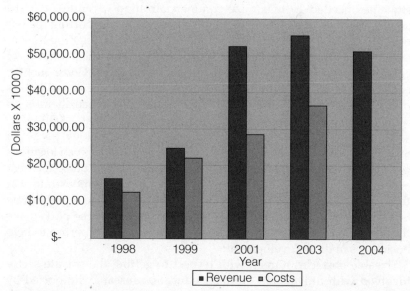

Source: Statistics Canada.

still active, resulting in a 79 per cent subsistence rate. Furthermore, 80 per cent of those companies still in business continued to be located in the same region as their associated university (Clayman and Holbrook 2003). Although this sample is relatively small, it does provide some evidence that the bigger Canadian universities have been successful at starting companies that survive for more than just a few years.

How Can the Diffusion of University Research Be Improved?

There are various commercialization strategies that may be incorporated to provide greater stimuli to the entrepreneurial environment at universities. For example, the current trend in licensing revenue from universities and hospitals has not reached optimal capacity. In order to address this issue, the federal government increased funding to the tri-councils (National Science and Engineering Research Council, Social Sciences and Humanities Research Council, and Canadian Institute of Health Research) while creating intellectual property management offices. The federal government has also had for decades government programs such as the Industrial Research Assistance Program (IRAP).

There has been much discussion concerning the different types of strategies needed to maximize commercialization revenues. For example, Feldman et al. (2002) highlight three factors supporting the use of equity stakes: opportunities for long-term rewards, aligning the objectives of both parties (licensee and licensor), and the branding of entrepreneurial flexibility to the external environment. Consequently, equity stakes are not without difficulties and there are normally disputes between the diverse stakeholders, especially when there is no dominant player in the ownership structure (Lockett et al. 2003).

An additional strategy could be to develop technology parks. In Sweden, studies have shown that firms that are based at such locations put greater weight on access to equipment, R&D, and personnel categories. Furthermore, the findings of these studies demonstrate that those companies that were based in the science park had a higher rate of job creation (Lofsten and Lindolef 2002). Overall, these parks have the potential to provide a wide range of services and expertise to help businesses in their development.

Sizeable obstacles remain with respect to getting the private sector involved with universities. Tertiary education research supported by the private sector has grown at a slower rate than that supported by the provincial and federal governments and this could reflect lower expectations with regards to perceived returns in productivity or product innovation. There has been some research suggesting that the role and potential of academia is misunderstood by the corporate sector. Thursby and Thursby (2003), for example, have highlighted some of the factors limiting the willingness of corporations to create research partnerships with universities in the United States.

The research consisted of a sample of 199 (of a possible 300) companies that had not licensed technology. The reasons given were the following: university research is generally at too early a stage of development (49 per cent); universities rarely engage in research in our line of business (37.4 per cent); universities refuse to transfer ownership to our company (31 per cent); university policies regarding delay of publication are too strict (20 per cent); and our company has concerns about obtaining faculty cooperation for further development of the technology (16 per cent). Zieminski and Warda (1999) conducted a similar study using a 1 (not important) to 5 (critical) scale, for which the top three results of 'why industry did not partake in partnerships' and 'what universities perceived as barriers for business' are listed in the table below:

University viewed barriers for industry		Industry viewed barriers for successful partnerships	
1 Poor knowledge of industrial needs	4.0	1 Shortage of R&D funds	3.7
2 Restrictive IP policies	3.9	2 Lack of partnership champions	3.6
3 Lack of relevant expertise	3.4	3 Emphasis on product development	3.5
3 Inability to adhere to agreed deadlines	3.4	3 Compressed development cycles	3.5

The reasons given are diverse and suggest that in any event much has to be done before a climate of cooperation and commercialization can be achieved. In addition to the structural reasons, different cultures and low mutual trust appear to be impeding progress. This suggests that these problems will not be quickly resolved and that educators, government, and industry need to foster better relationships between sectors to enable success to occur.

When considering the diffusion of university research, it should not be forgotten that the Canadian higher education sector has two main players – universities and community colleges/institutes – and both are shifting resources towards innovation and greater commercialization. A further issue to consider is how these two players can best ensure the development of complementary strategies and an optimal mix for their research mandates? Due to its relatively small population, Canada needs to harvest the maximum benefits from its current investment in research to contend with its international competitors. Consequently, any excessive duplication of roles, mandates, and strategies will hamper Canada's ability to successfully foster the innovation required to create the next generation of products and services.

This adds a further level of complexity in that, compared to universities, relatively little is known about community colleges and their applied research mandate. Currently there are approximately 200 colleges, of which the Association of Canadian Community Colleges represents 157 (Madder 2005; Corkery 2002). These educational institutions are involved with nearly 900 communities throughout Canada, fostering an understanding of business needs in many large and small localities (Madder 2005). Initially, a majority of community colleges were founded in the 1960s to develop the skill requirements of a manufacturing and service economy (Ivany 2000). Today's community colleges have a more expansive mandate of vocational training, specialized training, university transfer and preparation, and economic and cultural development (Skolnik 2004). Recently, the governments of Ontario and Alberta authorized the granting of four-year applied degrees in certain disciplines (ACCC 2006; Skolnik 2005). This legislation provided a competitive branding tool to compete with other institutions, including universities.

Since the release of the federal government's innovation agenda and due to the resounding popularity of this agenda with lobby groups requesting additional public funds, the community colleges have been attempting to further consolidate themselves in the field of applied research. In a study published by the ACCC in 2006, these institutions foresee themselves expanding in two main areas:

Development	*Commercialization*
1 Proof of concept	1 Occurrences in business and industry
2 Applied research	2 Product launch
3 Prototyping/simulation	3 New business start-up
4 Testing/analysis	4 Business development
5 Industrial/field/clinical trials	5 Business expansion
	6 Trade exploration
	7 Implementation of new policies

As of late, community colleges have had some success in garnering funds from the Canadian Foundation of Innovation ($28 million); and in 2004, NSERC created a College and Community Innovation Pilot Program to which colleges submitted 31 project applications for six funding slots (Madder 2005). The NSERC program was founded to promote factors that would nurture innovation by fostering the development of new products and services (NSERC 2007). Moreover, since 1983 the Quebec government has established and supported numerous College Centres for the Transfer of Technologies affiliated with CEGEPs to solve regional industrial issues (ACCC 2006; Belanger et al. 2005). Significantly, a study by Belanger et al. (2005) examined different enabling factors that fostered the development of this new applied research agenda. The four most desirable outcomes listed are: students' participating in R&D, increased status vis-à-vis other colleges, increased status vis-à-vis other non-university institutions, and more emphasis on economic development.

No country can support gross inefficiencies in its research policy by duplicating infrastructures in the university and the community college systems. On the other hand, it is foolhardy to believe that only universities should concentrate on research given the potential and rapid growth of colleges, research parks, incubation centres, and other commercialization infrastructure that are creating a much greater capacity for the development component of R&D. It is also important to acknowledge the current abilities and expertise of regional and local business connections available through community colleges. Consequently, we agree with the recommendation by Belanger et al. (2005) that a 'joint research only' fund be created to bring the different parties together through reciprocal partnerships that exploit synergies and knowledge transfer. Such collaboration would allow for a more detailed understanding of respective abilities and strengths and provide a building block towards collaboration and innovation.

Conclusions

Overall, if one looks at the enormous increase in higher education (mainly university) research funding, the expected commercialization revenue windfall and other related commercialization outputs have not been achieved. However, while the chapter sets out the case for improving the innovation and commercial application of university research, we believe it would be a misguided policy to do this at the expense of pure research. In particular, it would be counterproductive to concentrate solely on short-term gains while not attempting to develop ingenious discoveries that may be beneficial for the future. Andrew Sharpe, Executive director of the Centre for the Study of Living Standards, for example, makes this point in relation to medical research grants:

> Those research grants do not lead to short-to-medium term productivity growth because they do not translate instantly into money-making propositions for private firms. But academic funding from the research councils is really a major source of long-run productivity growth. (quoted in the *Globe and Mail*, 23 November 2006, A20)

Anderson also makes the related point that 'neglecting basic research or failing to distinguish it from applied research compromises universities' advantage over corporate R&D' (2001, 240). In the longer term, basic or pure research is crucial for applied research and innovation.

However, as Duderstadt (2005) points out, the reality is that 'public higher education institutions can no longer assume that public policies and investment will shield them from market competition' (2005, 85). Chapter 5 suggested that the CFI and other semi-autonomous agencies have protected Canadian universities from the full extent of market forces; however, in Canada, as elsewhere, pressures for greater commercialization reflect long-term trends in higher education, and the signs are that under the current Conservative government this trend will be intensified and accelerated. In his May 2006 Budget, Finance Minister Flaherty earmarked only $100 million for research and development, with only $40 million being added to the budgets of the three major federal granting councils: the Natural Sciences and Engineering Research Council of Canada, the Social Sciences and Humanities Research Council of Canada, and the Canadian Institutes of Health Research. After a decade of significant growth in research funding this has been seen by critics as a retrograde step. The *Globe and Mail* article cited above, for example, expresses fears that in the short term it will jeopardize the progress of researchers who are on the verge of many breakthroughs in areas such as cancer treatment. In the longer term, the *Globe*'s editorial board argues that the real test of Mr Flaherty's success will come when he reveals just how much importance he attaches to government-funded research, including the federal granting councils:

> That's not catchy. It's not a vote grabber. But that vital research does so much for Canadians' quality of life and standard of living. And so far Mr. Flaherty has been relatively miserly in his investments and his commitments. (*Globe and Mail*, 23 November 2006, A20)

Such criticism serves as a powerful reminder that commercialization and productivity, while desirable outcomes, are not the only criteria to consider in respect of funding research. Research, and the technology and products they spawn, can make an invaluable contribution to the quality and longevity of people's lives. In addition, if economic performance becomes the only criterion for funding, then it distorts the production of knowledge in ways that may have a destabilizing and a deleterious impact on societies in the longer term. One of Clark Kerr's pathologies for the modern university, predicted in 1963, was the 'elevation of the sciences above the humanities and the social sciences, creating a widening gap between the "hard" and "soft" sides of intellectual life, and between the "rich" and the "not so rich" participants' (Kerr 1963, 199). Though philosophers, political scientists, and sociologists may not make as direct or as tangible an impact on Gross National

Product as biotechnology, there is no question that they make an important contribution to society in general. Moreover, the interdisciplinary programs these subjects have spawned, such as MBAs and MPAs, provide leadership, policy, and entrepreneurial skills that are crucial for identifying, supporting, and commodifying scientific innovation.

Consequently, although we advocate greater state funding and resources for commercially directed university and private research, ideally this should not be done at the expense of other areas of research. In addition to the reasons stated in this chapter and elsewhere in the book, the animosity that can result from perceived unfairness in funding policies can seriously undermine interdisciplinary collaboration and results. The challenge for the government and for university stakeholders is to use financial incentives in a way that ensures a balanced approach to research and maintains a healthy equilibrium between academic disciplines and among research, teaching, and commercial development.

In addition, as the chapter has sought to demonstrate, the government's role in encouraging and facilitating applied research should not be limited simply to increasing funding. Federal governments in particular have a crucial role to play in creating the 'winning conditions' whereby commercialization can take place with minimal bureaucracy and adequate protections and incentives in place for individuals and their institutions. To this end, a clear and concise national intellectual property policy similar to the Bayh-Doyle Act in the U.S., giving universities full of control of discoveries, would help to stimulate entrepreneurial behaviour and foster commercialization revenues and opportunities.

The federal government is also uniquely placed to promote communication between regions, universities, and sectors, and to this end should consider a strategy for further emphasizing the integration of university research into SMEs with the creation of programs similar to the Small Business Innovation Research (SBIR) and the Small Business Technology Transfer (STTR) in the United States. Such a program would be similar to the Canadian SME Partnerships Initiative recommendations of the Industry Canada report 'People and Excellence: The Heart of Successful Commercialization.' This funding would have two components: 'feasibility funding,' creating limited funds for short-term investigation of scientific merit and feasibility; and 'prototype funding,' providing more extensive funds for the further development of projects with strong scientific and commercial merit (Industry Canada 2006, 29). This could be overseen by the creation of a Commercialization Partnership Board comprised of experienced public and private sector members and instituted to bridge the apparent disconnect between business and universities (Industry Canada 2006).

Drawing on national and international examples, the federal government can also play a significant role in supporting those universities that are successful in commercialization, while attempting to emulate these environmental contexts elsewhere in Canada. The University of Waterloo, for example, has been able to accomplish many firsts in the knowledge-focused economy and provides a made-in-Canada role model from which other institutions can learn. This university created one of Canada's largest research technology parks, which has been growing with a diverse group of tenants and services such as an Accelerator Centre to expand and develop new ventures. One of the University of Waterloo's greatest affiliations is with Research in Motion and its existence near campus. Other companies, such as Sybase Inc. and Open Text, have emerged as proven successes, while Google has made a purchase to participate in the regional benefits (Evans 2006). The economic future for this growing technology community is vibrant and consistent with the federal government's vision of how to maintain Canada's competitiveness.

It is doubtful that all Canadian universities could or even should adopt such a radical model; however, there are lessons for all Canadian universities and colleges as they each seek to exploit the commercial potential of their research. In addition to increased funding, a more entrepreneurial approach will demand a major shift in attitude by all stakeholders and require traditional mindsets and roles, such as the CFI's regulatory role discussed in the previous chapter, to be reconfigured in order to focus collaboratively on sustaining Canada's economic future, on which all Canadian universities ultimately depend. In this context the barriers to innovation, commercialization, and entrepreneurship are significant but, given global international competition and the pressures on public funding, unless these are addressed, Canadians generally, and future generations of students in particular, will likely pay a heavy price.

REFERENCES

ACCC. (2006). *Applied Research at Canadian Colleges and Institutes*. Association of Canadian Community Colleges.

Advisory Council on Science and Technology. (1999). *Public Investments in University Research: Reaping the Benefits.* Ottawa: Industry Canada.

Anderson, Melissa S. (2001). 'The Complex Relations between the Academy and Industry: Views from the Literature.' *The Journal of Higher Education* 72 (2): 226–46.

Audretsch, David B. (2003). 'Standing on the Shoulders of Midgets: The U.S. Small Business Innovation Research Program (SBIR).' *Small Business Economics* 20: 129–35.

Audretsch, David B., and Maryann P. Feldman. (1996). 'R&D Spillovers and the Geography of Innovation and Production.' *The American Economic Review* 86 (3): 630–9.

Barr, Nicholas. (2004). 'Higher Education Funding.' *Oxford Review of Economic Policy* 20 (2): 264–83.

Belanger, Charles H., Joan Mount, Paul Madgett, and Ivan Filion. (2005). 'National Innovation and the Role of the College Sector.' *The Canadian Journal of Higher Education* 25 (2): 27–48.

Bercovitz, Janet, Maryann Feldman, Irwing Feller, and Richard Burton. (2001). 'Oganizational Structure as a Determinant of Academic Patent and Licensing Behavior: An Exploratory Study of Duke, Johns Hopkins and Pennsylvania State Universities.' *Journal of Technology Transfer* 26: 21–35.

Bordt, Michael, and Cathy Read. (1999). *Survey of Intellectual Property Commercialization in the Higher Education Sector, 1998.* Ottawa: Statistics Canada.

Brzustowski, T.A. (2006a). 'Innovation = Invention + Commercialization: A Systems Perspective.' *Optimum Online*, 36 (3) (September): 1–8.

– (2006b). 'Innovation in Canada: Learning from the Top 100 R&D Spenders.' *Optimum Online*, 36 (4) (December): 1–9.

Bygrave, W.D., and C.W. Hofer. (1991). 'Theorizing about Entrepreneurship.' *Entrepreneurship Theory and Practice* 16 (2): 13–22.

Cameron, David M. (2002). 'The Challenge of Change: Canadian Universities in the 21st Century.' *Canadian Public Administration* 45 (2): 145–74.

Caniels, Marjolein C.J., and Henny A. Romjim. (2003). 'SME Clusters, Acquisition of Technological Capabilities, and Development, Concepts, Practice, and Policy Lessons.' *Journal of Industry, Competition and Trade* 3 (3): 187–210.

CAUT. (2006). *CAUT Almanac of Post-Secondary Education in Canada.* Available from: www.caut.ca (cited 1 November 2005).

Clayman, B.P., and J.A. Holbrook. (2003). *The Survival of University Spin-offs and Their Relevance to Regional Development.* Canada Foundation for Innovation.

Corkery, K. (2002a). *Colleges and the National Innovation Agenda.* Ottawa: Strategic Policy Branch, Industry Canada, Government of Canada.

Davenport, Paul. (2002). 'Universities in the Knowledge Economy.' In *Renovating the Ivory Tower*, ed. David Laidler, 41–59. Toronto: C.D. Howe Institute.

Duderstadt, James J. (2005). 'The Future of Higher Education in the Knowledge-Driven Global Economy of the Twenty-first Century.' In *Creating Knowledge, Strengthening Nations: The Changing Role of Higher Education*, ed. Glen A. Jones, Patricia L. McCarney, and Michael L. Skolnik, 81–97. Toronto: University of Toronto Press.

Evans, Mark. (2006). 'The "Quiet Boom": High-Tech Turnaround Drives Ottawa, Waterloo Growth.' *Financial Post*, January 31. Available from: http://www.canada.com/nationalpost/financialpost/story.html?id=2a0b745b-eded-4269-af0f-f454d25a8ca6&p=3 (cited 4 October 2006).

Feldman, Maryann P. (1999). 'The New Economics of Innovation, Spillovers and Agglomeration: A Review of Empirical Studies.' *Economics of Innovation and New Technology* 8 (1&2): 5–25.

Feldman, Maryann, Irwin Feller, Janet Bercovitz, and Richard Burton. (2002). 'Equity and the Technology Transfer Strategies of American Research Universities.' *Management Science* 48 (1): 105–21.

Fitagawa, Fumi. (2005). 'Constructing Advantage in the Knowledge Society Roles of Universities Reconsidered: The Case of Japan.' *Higher Education Management and Policy* 17 (1): 1–18.

Fisher, D., and J. Atkinson-Grosjean. (2002). 'Brokers on the Boundary: Academy–Industry Liaison in Canadian Universities.' *Higher Education* 44: 449–67.

Fritsch, Michael, and Christian Schwirten. (1999). 'Enterprise-University Co-operation and the Role of Public Research Institutions in Regional Innovation Systems.' *Industry and Innovation* 6 (1): 69–83.

Gibbons, M. (1998). *Higher Education Relevance in the 21st Century*. UNESCO World Conference on Higher Education.

Grandi, Alessandro, and Rosa Gimaldi. (2005). 'Academics' Organizational Characteristics and the Generation of Successful Business Ideas.' *Journal of Business Venturing* 20: 821–45.

Guthrie, Brian. (2006). *Picking a Path to Prosperity: A Strategy for Global-Best Commerce*. The Conference Board of Canada.

Henrekson, Magnus, and Nathan Rosenberg. (2001). 'Designing Efficient Institutions for Science-Based Entrepreneurship: Lessons from the U.S. and Sweden.' *Journal of Technology Transfer* 26: 207–31.

Hum, Derek. (2000). 'Reflection on Commercializing University Research.' *The Canadian Journal of Higher Education* 30 (3): 113–26.

Industry Canada. (2006). *People and Excellence: The Heart of Successful Commercialization*. Ottawa: Public Works and Government Services Canada.

Ivany, R. (2000). 'Economic development & a new millennium mandate for Canada's community colleges.' *College Canada* 5 (1): 10–13.
Kerr, Clark. (2001). *The Uses of the University*. 5th ed. Cambridge, Mass.: Harvard University Press.
Laidler, David. (2002). 'Renovating the Ivory Tower: An Introductory Essay.' In *Renovating the Ivory Tower*, ed. David Laidler, 1–38. Toronto: C.D. Howe Institute.
Langford, C.H., J. Hall, Peter Josty, Stelvia Matos, and Astrid Jacobson. (2006). 'Indicators and Outcomes of Canadian University Research: Proxies Becoming Goals?' *Research Policy* 35: 1586–98.
Lockett, Andy, Mike Wright, and Stephen Franklin. (2003). 'Technology Transfer and Universities' Spin-Out Strategies.' *Small Business Economics* 20: 185–200.
Lofsten, H., and P. Lindelof. (2002). 'Science Parks and the Growth of New Technology-Based Firms.' *Research Policy* 31 (6) (August): 859–76.
Lloyd-Ellis, Huw, and Joanne Roberts. (2000). *Twin Engines of Growth*. Available at: http://www.chass.utoronto.ca/ecipa/archive/UT-ECIPA-JOROB-00-02.pdf (cited 20 April 2006).
Madder, D.J. (2005). *Innovation at Canadian Colleges and Institutes*. Association of Canadian Community Colleges.
Newman, F., L. Couturier, and J. Scurry. (2004). *The Future of Higher Education Rhetoric, Reality, and the Risks of the Market*. San-Francisco, Calif.: Jossey-Bas.
Noftsinger, Jr, John B. (2002). 'Facilitating Economic Development Through Strategic Alliances.' *New Directions for Higher Education* 120: 19–28.
NSERC. (2007). 'College and Community Innovation Pilot Program.' Natural Sciences and Engineering Research Council of Canada. Available from: http://www.nserc.gc.ca/colleges/college_desc_e.asp (accessed 11 January 2009).
OECD. (2003). *Knowledge Management: New Challenges for Education Research*. Organization for Economic Co-operation and Development.
Park, Walter G. (1994). 'Adoption, Diffusion, and Public R&D.' *Journal of Economics and Finance* 18 (1): 101–23.
Powers, Joshua B. (2004). 'R&D Funding Sources and University Technology Transfer: What is Stimulating Universities to be More Entrepreneurial?' *Research in Higher Education* 45 (1): 1–23.
Powers, Joshua B., and Patricia P. McDougall. (2005). 'University Start-up Formation and Technology Licensing with Firms that Go Public: A Resource-Based View of Academic Entrepreneurship.' *Journal of Business Venturing* 20: 291–311.
Pries, F., and P. Guild. (2007). 'Commercial Exploitation of New Technologies Arising from University Research: Start-ups and Markets for Technology.' *R&D Management* 37 (4) 319–28.

Read, Cathy. (2000). *Survey of Intellectual Property Commercialization in the Higher Education Sector, 1999.* Ottawa: Statistics Canada.
– (2003). *Survey of Intellectual Property Commercialization in the Higher Education Sector, 2001.* Ottawa: Statistics Canada.
– (2004). *Survey of Intellectual Property Commercialization in the Higher Education Sector, 2003.* Ottawa: Statistics Canada.
Siegal, Donald S., David Waldman, and Albert Link. (2003). 'Assessing the Impact of Organizational Practices on the Relative Productivity of University Technology Transfer Offices: An Exploratory Study.' *Research Policy* 32: 27–48.
Simha, Robert O. (2005). 'The Economic Impact of Eight Research Universities on the Boston Region.' *Tertiary Education and Management* 11: 269–78.
Skolnik, Michael. (2004). 'The Relationship of the Community College to Other Providers of Postsecondary and Adult Education in Canada and Implications for Policy.' *Higher Education Perspectives* 1 (1): 36–58.
– (2005). 'Reflections on the Difficulty of Balancing the University's Economic and Non-Economic Objectives in Periods When Its Economic Role is Highly Valued.' In *Creating Knowledge, Strengthening Nations: The Changing Role of Higher Education,* ed. Glen A. Jones, Patricia L. McCarney, and Michael L. Skolnik, 106–26. Toronto: University of Toronto Press.
Slaughter, S., and L.L. Leslie. (1997). *Academic Capitalism: Politics, Policies, and the Entrepreneurial University.* Baltimore: Johns Hopkins University Press.
Statistics Canada. (1997). *Commercialization of Intellectual Property in the Higher Education Sector: A Feasibility Study.* Ottawa: Statistics Canada.
– (2004). *Estimation of Research and Development Expenditures in the Higher Education Sector, 2002–2003.* Ottawa: Statistics Canada.
– (2008). *Research and Development Expenditures.* Ottawa: Statistics Canada.
Thompson, Janet. (n.d.). *Estimates of Canadian Research and Development Expenditures (GERD), Canada, 1994–2005 and by Province 1994–2003.* Ottawa: Statistics Canada.
Thursby, Jerry G., and Marie C. Thursby. (2000). *Industry Perspectives on Licensing University Technologies: Sources and Problems.* Available from: http://www.provendis.info/fileadmin/info/pdfs/1255.pdf (accessed 11 January 2009).
Thursby, Jerry G., and Sukanya Kemp. (2002). 'Growth and Productive Efficiency of University Intellectual Property Licensing.' *Research Policy* 31: 109–24.
Tödtling, Franz, and Alexander Kaufmann. (2002). 'SMEs in Regional Innovation Systems and the Role of Innovation Support: The Case of Upper Austria.' *Journal of Technology Transfer* 27: 15–26.

Trajtenberg, Manuel. (2001). 'Government Support for Commercial R&D: Lesson from the Israel Experience.' NBER Conference on Innovation Policy and the Economy, Washington, D.C., April.

Youl Lee, S., R. Florida, and Z.J. Acs. (2004). 'Creativity and Entrepreneurship: A Regional Analysis of New Firm Formation.' *Regional Studies* 38 (8): 879–91.

Zieminski, Janusz, and Jacek Warda. (1999). *Paths to Commercialization: University Collaboration Research.* Conference Board of Canada.

7 Intellectual Property, Technology Offices, and Political Capital: Canadian Universities in the Innovation Era

MALCOLM G. BIRD

This chapter builds on the broader discussion of commercialization as discussed in chapter 6 by Madgett and Stoney, but focuses more particularly on how Canadian universities are approaching issues of intellectual property rights, patents in particular, and in developing technology transfer offices (TTOs) as features of managing their new roles in the innovation era and the 'knowledge-based' economy.[1] The federal influence via intellectual property law is direct, whereas TTOs emanate from universities' own judgment about how to respond to the varied federal and provincial pressures to disseminate the research and knowledge they produce. The intellectual property (IP) generated by universities, particularly patents, and the role of TTOs in disseminating knowledge are two key factors that can not only help to improve the effectiveness and public image of these institutions in knowledge transfer and innovation, but also contribute to their aggregate amount of political capital. Political capital refers here to the universities' continuing need to maintain overall political and funding support and political legitimacy in a world where governments have an extensive number of competing demands for scarce public resources.

---

1 This chapter is based on published reports and documentation, and on a number of discussions with university officials who are familiar with technology transfer processes. Special thanks are owed to the participants from Carleton University, University of Ottawa, Waterloo University, University of British Columbia, and Queen's University. The universities selected for more illustrative analysis were chosen to capture some regional factors and issues of university size and types of TTO, but they do not constitute a full representative sample. Waterloo, in particular, was chosen because of its eminence in a technology cluster in south-central Ontario. Participant universities, however, had diverse ownership and royalty rules regarding IP.

This chapter examines the importance of using patents to solidify specific property rights, and demonstrates that patenting assists universities in their efforts to disseminate knowledge. It also looks at some data on patents and Canadian universities, exploring which universities are filing patents and outlining the aggregate value of the royalties accrued.

TTOs fulfil a broader knowledge dissemination role by helping to transfer university-generated ideas to the external world. They are, in short, interfacing or boundary agencies, as discussed in chapter 1. TTOs have become part of the structure of many universities, in part because universities are looking for alternative funding sources (to earn income from IP licensing) principally by commercializing university-generated knowledge in the marketplace, but also to publicize such commercialization and knowledge transfer as a tangible measurement of university outputs. TTOs also disseminate university-generated technology to facilitate the creation of new firms or expand existing ones. In doing so, they help to create high-calibre technology-based jobs that are valuable to communities. It is the role, however, that universities play in a more general sense, in providing skilled graduates and generating new ideas, that makes them such important institutions for diversifying the Canadian economy and bringing our country into the innovation age.

This chapter will argue that in light of the intense pressure for scarce public funding, the shift to a 'knowledge-based' economy, and the pivotal role that universities play in the new innovation and also 'informational' age, universities must attract public support. Creating and maintaining a positive institutional image in the eyes of the public, and especially the political elite, is one method of ensuring significant and sustained overall public funding for Canadian universities. Universities, like all other public institutions, need to leverage their political capital in order to maximize public expenditures.

The increasing financial pressure on governments to provide public goods and the structural shift towards an economy characterized by information, innovation, and knowledge has had a profound effect on the operation and role of Canadian universities. Many have expanded their mission and, either explicitly or implicitly, have extended their seminal role as institutions of teaching and research into the realm of economic development as well. The 'entrepreneurial' university, as chapter 6 has shown, focuses on creating linkages with the external world, especially the marketplace, not only by providing qualified graduates and conducting basic research, but also by making efforts to commercialize its technical discoveries.

This shift has occurred because of both internal and external influences on universities. It is specifically intended to improve the economic performance of the nation or region, the financial performance of the university itself, and the financial performance of its faculty (Etzkowitz et al. 2000, 313). Within the Canadian context, improving the public image of universities and augmenting the perception, especially among the political elite, that universities are key institutions for ensuring a strong, diversified Canadian economy is another critical reason for this shift. Improving Canada's 'asset mix' and increasing the amount of government funding for university research so that Canada can secure its position as a player in the knowledge-based global economy is how David Naylor, President of the University of Toronto, aptly argued for continued government funding (quoted in Crane 2006).

There are, however, many who oppose universities' increasingly entrepreneurial approach, and their increasingly 'corporate' characteristics, viewing such changes as a grave threat to the independence of universities (Turk 2000; Tudiver 1999). While their concerns have some validity, the needs of universities to appeal to the requirements of elected governments, and the desires of these governments, to use universities, or to see them function as key economic growth institutions, often appear to override such trepidations. Having said this, universities must nevertheless continue to strive to be autonomous institutions, operating at arm's length from the government, and critics need to have some faith that they will be able to mitigate the direct political pressures attempting to influence how they function, while still remembering the needs of democratic governments; after all, it seems logical that he who pays the piper, as it were, should have some influence on the tune played. Striking a balance between these two directives – autonomy and demonstrable relevance – is of seminal importance in understanding the changing federal role vis-à-vis universities, and provincial roles as well.

Chapters 1, 2, and 3 provide an outline of the contemporary context of Canada's universities and of key federal policy and funding regarding research and innovation. For our purposes, however, it is important to further illustrate two related points. First, the significant funding increases noted in chapter 1 and in other studies (AUCC 2005, 2 and 15) have had a profoundly positive impact on the quality and quantity of research being conducted, and thus universities are eager to ensure that the additional funds continue (Challis et al. 2005, 360–1). Universities are in a competitive political environment, in which vital sectors such as health, among many others, are fighting over limited public funds. They recognize the need to ensure that governments understand the value of their contributions to the economy and to society (Iacobucci and Tuohy 2005, xvii).

Second, the role of the university as a vehicle of economic development is actually quite consistent with the historical development of universities in the post-war period (Axelrod 1982), and also earlier when Canada's economy was a resource-based economy. More generally, it is indicative of the key role that the state has played (and continues to play) in Canadian economic and social development (Hardin 1974).

The structure of the chapter is a quite straightforward one. The first section examines intellectual property issues and responses, and the second section examines the role and function of the TTO. Conclusions then follow.

IP and Canadian Universities: Dissemination versus Protection

Intellectual Property (IP) is, very generally, comprised of four main groupings: trademarks, copyrights, design rights, and patents. For the purpose of this chapter, we are principally concerned with patents, which are, very simply, a monopoly right on a specific technological invention that can be a physical apparatus, method, or particular process, for a given period of time (Doern and Sharaput 2000, 18–19). The holder of the patent has a state-enforced monopoly on the production and use of the apparatus, method, or process, and he or she often earns royalties or licensing fees from it. Copyright, too, is relevant to universities, but the author owns most academic production and the possibility for royalties from most academic work is small. The other two types of IP, design rights and trademarks, are less relevant to our discussion.

Before going further, an important issue must be quickly examined, because, as will be explained, it has particular relevance to universities and their position among public agencies. This issue is the basic trade-off between knowledge protection and dissemination under IP provisions regarding patents. IP systems of law are intended both to protect inventors by conferring a monopoly right, and to disseminate knowledge because patents are made public. Thus the knowledge embodied in the patent is put in the public domain so others might eventually and indirectly improve on it further. Patents in this sense involve the provision of both public goods and private goods.

All information, to a large extent, is a public good. Public goods, very broadly, are defined by two characteristics; they are non-rivalrous with respect to their consumption (in other words, one's consumption of the good does not impinge upon the rights of others to likewise consume it), and non-exclusive (meaning that once the good is produced, it is difficult, if not impossible, to prevent someone from enjoying it). National defence, public parks, and public broadcasting are typical examples of public goods (Friedman 2002, 96). In the case of valuable information or inventions, their 'public good' characteristics would allow anyone to take advantage of their usefulness for his or her own purposes, unless some type of protection against use were to be afforded to the creator or inventor. A creator would have little incentive to invest time and energy into an idea without such protection, since the benefits that would flow to compensate him or her for his or her efforts would be minimal. This is why the state intervenes. It grants patents to producers (monopoly rights), which are enforceable through the courts. But at the same time, the patent is made public. The more information is publicly available for use by everyone, the more productive and efficient industry, and indeed, society as a whole, becomes. The desire to encourage the development of ideas by providing incentives (in the form of property rights protection) to individual creators, so that they will continue creating, must be weighted against the public interest in maximizing the amount of information available to individuals, in order to minimize inefficiencies (Doern and Sharaput 2000, 19).

Once protection is granted under a patent, individuals or firms are able to access the patented item's unique properties. But if they want to benefit commercially from the patent per se, there is a price to be paid. Before allowing others to benefit commercially, the patent owner usually demands some form of royalty or rent. A university, naturally, should seek to disseminate knowledge as much as possible, and patenting knowledge is one essential manner in which they can go about increasing the supply of knowledge and information. Plentiful (or perfect) information, after all, is one of the tenets of achieving optimal output.

IP and patent rights, however, vary from country to country. In Canada and the United States, for instance, they are well protected, and a firm (or university) that owns a patent is relatively sure that their rights are enforceable by courts and that there is a general respect for all types of IP. Existing pan-national agencies are not fully capable of enforcing IP rights throughout the world, and thus patent holders are dependent on national governments to regulate the systems that protect their particular property right from encroachment. For the owners of patents this system works well, as long as all countries enforce their IP laws in the same ways as their North American counterparts and as long as the patent holder can afford the costs of legal action.

Things, however, are very different in China. Around Shanghai there is a saying, 'We can copy everything but your mother,' that ought to send chills up any patent holder's spine. And copy they do. Almost every conceivable item can be found in a pirated form in China, including the illegitimate reproduction of entire automobiles (*Economist* 2007). Such a lack of respect for IP might not be so problematic except for two critical facts: first, that China is actively seeking to move up the value-added manufacturing ladder and is (and will be) making many high-end (and complicated) devices in the future; and second, that China's lack of enforcement of IP rights is a deliberate set of inactions whose purpose is to help provide Chinese firms with knowledge from the rest of the world for free (Fishman 2006, 231–52). As world communication and transportation networks grow more efficient and countries such as Canada rush to join the knowledge-based economy, we will face some stiff competition from countries like China. It will be competition, furthermore, that does not always follow one set of property right ideals. Firms – and especially universities – that produce IP and seek to commercialize it need to be well aware that their property rights, in a global market, are not as secure as is often thought.

Canada's federal patent law does not have a counterpart to America's Bayh-Dole Act of 1980. This is an Act, very simply, that specifically establishes that the ownership of IP derived from federally funded research lies with the university and inventor, not with the federal government (Leaf 2005). In the absence of such a Canadian federal equivalent, IP ownership policies at Canadian universities are diverse.

In general, there are three types: inventor ownership, university ownership, or a combination of the two, where ownership rights are arrived at through use of a predetermined set of criteria. At Carleton University, for example, the inventor owns the IP. In fact, his or her right to do so is enshrined in the faculty's collective agreement. The same goes for Queen's University and the University of Waterloo. In contrast, at the University of Ottawa the opposite is true, and ownership resides with the university. Inventor-owned advocates argue that this model encourages entrepreneurship among faculty, who, spurred on by a possible financial benefit, will gladly invest energy and resources in a project or idea to bring it to fruition. This incentive, they argue, is particularly important in the early stages of a project, where the future benefit may not always be clear; under such circumstances it is all the more important to have the prospect of a big pay-off for the inventor in the event of success.

Advocates of university-owned patents argue that, under their model, the process of IP ownership is more predicable, since the licensee only has to interact with one owner (the university), and there is less chance of uncertainty around proprietary rights as the product is developed (and royalties earned). There are some positive and negative aspects to each type of arrangement, and there are ample success stories supporting each of them as well (Brzustowski 2006, 3).

The increased pressure on universities to produce IP with commercial potential has come not only from the political sphere through federal exhortation, but also from the need to find alternative sources of income that are not from users (students) or the government. This is a direct result of the increasing financial pressure that all public sector organizations face, and, not surprisingly, is directly related to the increasing reluctance by many individuals (and firms) to pay taxes in support of publicly provided goods. One of the many effects of this pressure has been a movement in Canadian society generally, and in public policy more specifically, to encourage or require government agencies to adopt principles from the private sector and act more like businesses, or to adopt in varying degrees New Public Management principles (Pal 2001). Universities are not immune from these structural changes in political economy (Fischmann 2005, 4).

So, if such significant emphasis is currently placed on Canadian universities to produce IP, especially patents, which have commercial potential, how are Canadian universities doing? A study by Statistics Canada provides some valuable insights.[2] According to Statistics Canada, Canadian universities filed 166, 145, and 150 patents in the years 2003, 2004, and 2005, respectively. Ownership of these patents

---

2 A few important notes on the Statistics Canada report: First, it includes patents filed by hospitals and universities, and only 44 per cent of the former and 81 per cent of the latter participated in the survey. Second, it only includes two patents filed by the University of Waterloo over the three-year period. This is because UW's faculty and students hold exclusive rights to the IP that they produce. In addition, the report does not account for other types of IP ownership rights that exist at Canadian universities. Since IP often belongs to the creator, the report may underestimate the number of patents filed. Finally, the report does not include filings by formal organizations that may be affiliated with universities, unless the name of the organization has the name of the university included in it.

was very much skewed to the older, more established and larger, institutions, perhaps because these universities have a large array of faculties, including medicine and engineering. To provide a few examples: over the three-year period described above, the University of British Columbia (UBC) produced 82 patents; the University of Alberta, 34; the University of Ottawa, 15; Queen's University, 33; McGill, 56; Université Laval, 39; and Université de Montréal, 29. Interestingly, and providing a slight contradiction to the general trend, the University of Toronto, Canada's largest university, only produced 10 patents over this period (Statistics Canada 2006).

If one compares the above numbers to the numbers of patents produced by the generally smaller, newer schools, the extent of the skew becomes apparent. The University of Victoria, for example, produced only 2 patents over this period; the University of Regina, 1; Carleton University, 1; Université du Québec à Montréal, 4; and the University of Prince Edward Island, 1. Again, there are a few exceptions. For example, York University in Toronto, the country's second biggest university, and a university that focuses on the Arts and Humanities, filed 2 patent applications, while Dalhousie University in Halifax, an older, more established school, filed only 1 patent (Statistics Canada 2006). The total number of patents held by Canadian universities and hospitals in 2003 was 3,047, an increase of 43 per cent from two years earlier, and of these, 45 per cent were commercialized in some form or another (Statistics Canada 2006, 4).

Drawing on these examples is not, in any way, faulting those that filed fewer patents, but only serves to illustrate that using patent filings as a way to measure 'output' from the universities is problematic, since the numbers are positively correlated with other factors (age of school, size, types of departments, etc.). It does not, for example, account for the direct impact that a new discovery might have on people's everyday lives, nor does it indicate the level of active research. In the case of the latter, measuring monies generated from external sources (either private firms or government funding bodies) is often touted as a superior indicator, since it conveys the gross research activity conducted on a particular new discovery. These numbers also demonstrate that universities have divergent policies, particularly with respect to the role and function of the TTOs (see more below). These divergent policies, as will be explained, affect how universities view the dissemination (or protection) of IP.

The total amount of money earned from the licensing, assigning, or otherwise commercializing of IP in 2003 was $55.5 million. Considering that the annual expenditures for UBC in the same year were $1.069 billion, this is a paltry amount (UBC 2003, 32). In addition, it is important to note that a significant proportion of the funds earned comes from a few, very lucrative, patents, many of which are due to expire soon. Sherbrooke University, for example, has a patent on a mechanism that compresses voice data into a digital signal for cellular telephones that has been very lucrative for the school; this, however, is an exceptional type of patent (Bird 2006).

Developing and commercializing IP also has some significant costs to universities. In 2003, universities and hospitals in Canada spent $36.4 million managing their IP, up from $28.5 million in 2001 (Statistics Canada 2006, 2). These costs, naturally, must be accounted for when measuring total income that public institutions earn from IP. The pattern of a few schools filing the vast majority of patents and earning the majority of the IP income is similar to the pattern in the United States, where one third of the significant discoveries and over half of the income from commercialization of inventions/ideas are generated by ten schools, such as MIT and Stanford (Leaf 2005, 4).

Technology Transfer Offices

Most Canadian universities have some form of technology transfer office (TTO). The function of a TTO is to help bring university-generated IP, usually in the form of a patentable idea, to market. A TTO acts as an interface or boundary institution in that it tries to 'bridge' the gap between universities and the marketplace. It is difficult to make broad generalizations about TTOs, since each university has differing rules for IP rights, and different strategies for enforcing them.

The University of Ottawa's TTO will serve as an initial illustrative example for the purposes of this chapter. Its TTO attempts to link university-generated inventions to the needs of the marketplace; they attempt to 'package' the product so that a market actor (usually a firm) can develop, market, and (hopefully) derive revenues from it, revenues which both the university and the inventor will share.[3] Generally, the

---

3 Royalties from the vast majority of IP patents are small. However, in the case of the University of Ottawa, when they accrue, they are divided between the university and the inventor(s) according to a prescribed formula in the university's collective agreement with its faculty. For the first $100,000 in royalties, 80 per cent go to the researcher and 20 per cent to the university. Over $100,000 they are split 50/50. The researcher is free to take cash or to deposit the monies into a research account where the first $5,000 will be matched by the university.

goal is to get the technology into the market, where it can be adequately developed, rather than to develop, market, or manufacture the product 'in-house' or to create a spin-off company. Established firms have expertise and capital that most faculty or staff in the university environment do not possess; thus, most of the time they are seen as being better equipped to turn the idea or invention into something marketable.

There are other TTO models in operation. Queen's University's technology transfer office, PARTEQ, for example, is run as a non-profit university company. It has not only started private firms, but has run them as successful businesses before selling them to the private market. It is one of the more successful technology transfer offices, however, and is, in many respects, well beyond the norm when compared to other Canadian TTOs.

The University of Ottawa, on the other hand, has a policy whereby all patentable IP must be disclosed to the university by the inventor (a sort of 'right-of-first-refusal' for the university). Once an idea has been disclosed, and if the inventor and his or her team are interested in pursuing commercialization, the first task of the TTO is to assess the quality of the invention and, more importantly, to examine the market to ensure that it has some commercial feasibility. Next, it will assist the developer(s) to create a working prototype and/or help to present the idea in language that non-university people can easily comprehend. TTOs also often possess and provide a limited amount of seed money to assist with prototype creation. They decide what type of property protection is needed and will liaison with experts in obtaining patents or other types of protection. In addition, TTO personnel provide assistance in finding capital and in developing a specific business plan to get the particular product to the market.

The type of business plan and the particular process adopted will naturally depend on the type of product under examination. Knowledge of basic science, business, and clear writing and communication skills are the most important assets of TTO officers. Many TTOs, like the one at the University of Ottawa, are in an embryonic stage of development and are still perfecting their procedures. It is also critical to note that development procedures take significant amounts of resources and that product development is a long-term process, which often takes five to ten years.

Most TTOs are focused on bringing products to the market in order to share the value of the invention with the external world. This does not mean, however, that they are driven by financial gain, since very few products produce significant royalty returns for either the inventor or the university. Rather, it is hoped that bringing a particular product to the market will create jobs and revenue for the company, and that positive economic effects will be felt throughout the entire community. Furthermore, commercialization policies are also pursued at universities, and by TTOs in particular, because both levels of government and their funding bodies mandate that universities make efforts to commercialize their knowledge. The amount of commercialization is a relatively easy way to gauge the 'output' being created at publicly funded universities.

The universities, however, have other reasons for pursuing technology transfer programs through TTOs. These reasons have less to do with earning monies directly from the commercialization of a particular product, and more to do with improving the public image of the university. Products that individuals can purchase in the marketplace and that have a tangible effect on their lives, such as new medical drug treatments, can help a large number of ordinary Canadians by improving their day-to-day lives. This will, obviously, enhance the reputation of the university that brought this product into the public sphere.

A strong reputation will make attracting and retaining high-quality faculty to the university easier, and will also help to attract first-rate students. In addition, a strong reputation based, at least in part, on some commercialization, will aid the university which, in attracting donations, can be leveraged to help attract additional research monies, and, of course, will encourage government and industry to invest in capital projects, such as new facilities, at the university. Being able to show the public (and, by extension, the political elites) that university-generated knowledge can and often does have a widespread impact on society is a valuable residual effect of technology transfer.

Accordingly, once a TTO is no longer a cost centre for its governing university, it will often shift its focus to maximize aggregate positive effect on society, thus improving the overall image/reputation of the university, rather than focusing on profit maximization per se. University administrators are well aware of the political value of transferring knowledge to the external world.

This process of turning an invention or idea into something marketable is, however, fraught with problems. Many of these problems stem from the inherent structural differences between the operation of universities and their faculties, and the needs of firms in the marketplace. Without delving too far into this subject (for a more extensive treatment of it, see Bird 2006), a number of factors make transferring university knowledge to the marketplace difficult. One of the most important of these is that the needs of the market actors (often venture capitalists) do not mesh with those of the university's faculty or students. For example, inventors often need to provide a substantial amount of data to the developers, data that is often not of publishable quality, but which still requires a significant amount of energy and time to produce.

In addition, the length of time it can take for a firm to develop a product, and the firm's need to maintain secrecy, at least during the initial phase of development, is at odds with the needs of faculty and graduate students to either publish their findings or complete their degrees. Publishing articles, after all, is the most important activity in terms of acquiring tenure or augmenting one's CV in order to successfully pursue an academic career. Oftentimes, faculty resent the directions of either firms or venture capitalists who have acquired their IP material, even though such direction may be necessary in order to turn the idea into a successful commercial product. Finally, 'success' in the marketplace frequently has as much, if not more, to do with the quality of people bringing the product to market than with the quality of the product itself.

As the broader account of commercialization and entrepreneurship presented in chapter 6 has shown in more detail, there are other problems as well. For example, there is a lack of higher-risk venture capital and a dearth of quality business managers, both within the university environment (often because wages paid at technology transfer offices are not commensurable with rates needed to attract experienced product developers) and within the Canadian business community. This is in addition to the often commented upon less developed 'entrepreneurial' spirit found in Canada as compared with the United States, especially in areas such as Silicon Valley. Moreover, even when universities are successful in their quest to commercialize knowledge, the knowledge is often purchased by a foreign firm (usually American), with the result that the jobs, revenue, and the product itself are often exported. Bear in mind that the problems are mostly associated with the *commercialization* process of IP and not with the actual IP itself; therefore, it is the *interface* between the firms and university that is problematic. Commercialization is only one way to transfer knowledge outside of the universities and it is substantially more difficult than the most important method of moving information from the university to the external world, namely by 'a pair of shoes' (Bramwell and Wolfe 2005, 27) worn by university students hired by firms.

Despite being aware of the inherent structural problems involved in the commercialization of university-generated knowledge, universities, by and large, are not addressing these problems. This is partly due to the fact that universities are not designed, in an institutional manner, to commercialize knowledge, but also because their democratic decision-making structure, and the general diffusion of responsibility within them, is not conducive to imposing radical change. Radical change would be needed in order for them to adequately address these deficiencies. This fact, however, might not be a real detriment to universities since they are (and should remain) dedicated to teaching and research, and need to be highly protective of their institutional autonomy. It also might be because of a general reluctance to adhere to one standard measure of their 'output,' and the heated debate over the type and form of the metrics needed to account for the activity at a publicly funded university. In the absence of pan-Canadian benchmarks, there is no agreed upon way to measure the output from universities (Canadian Council on Learning, 2006). Universities themselves might not want to address the fundamental structural problems because it may simply be impossible for them to do so without departing from their teaching and research function, and without losing essential autonomy from the political sphere. They are happy to accept monies from governments, but might actively resent a more interventionist encroachment on their institution's turf.

Carleton University's TTO is of interest in comparative Canadian terms. It has a vision that differs significantly from the 'standard' disclosure–evaluation–patent–licence process. It illustrates that even a non-commercial TTO can have significant 'image-improving' value to a university. Its approach to technology transfer is largely due to Carleton's faculty-ownership IP policy, its smaller faculties of engineering and hard sciences, and the absence of programs in the medical sphere. Carleton's policy can be described as a more 'holistic' approach. It is focused on the general dissemination of a wide variety of knowledge (from the hard and social sciences) from the university to its external environment. In this approach, the university is a central part of the public infrastructure and a pivotal player in providing firms with capable, skilled employees. By producing such capable, skilled employees, and increasing the knowledge base of a city or region, the university plays a key role in attracting firms to the university's geographic area. The result is a 'cluster' of specialized individuals and firms that are connected by formal and informal interpersonal networks. This, according to some, is the most effective method of bridging the gap between the market and the university. The best example of this phenomenon at work in Canada is provided by the University of Waterloo, and the 'cluster' role it plays in south-central Ontario (Wolfe 2005; Bramwell and Wolfe 2005).

Carleton University's TTO operates internal and external internships to provide students with real-world experience with firms in their areas of expertise, and it actively seeks to expand the social networks of individuals in related industries. It also works to link the university to firms through research partnerships and offers support to individuals seeking to patent their IP by, for example, providing seed money to develop a prototype and assisting with the technical details of patent acquisition. However, it has no financial incentive to do so, preferring to rely on the 'goodwill' created by initial support during the preliminary phases of development. This 'goodwill' brings significant benefits to the university. For example, successful spin-off entrepreneurs (of which Carleton has produced a goodly number) will hire interns, donate money, and, more generally, continue to contribute to expanding the networks of individuals on the strength of that 'goodwill.'

Other TTOs, too, place significant emphasis on research partnerships through, for example, collaborative methods, but find quantifying their knowledge transfer value, especially to politicians and senior civil servants, quite difficult. One of the advantages to using commercialization to disseminate university-produced IP is that it is a process that is easy to understand and one that many people value. When successful, it has tangible, measurable results.

Most interestingly, the Carleton TTO does not limit its support to the sciences or engineering faculties. One example demonstrates this well. David Carment, a professor in the Norman Patterson School of International Affairs and an expert in international conflict, has used some seed money from the Foundry Program to start his Internet-based initiative entitled the Snowball Project. Through an Internet portal, Carment and his team of researchers have compiled a wide range of information on potential conflict areas. The content of the portal ranges from the very broad, with information on the economic, political, and security environment contained in reports and briefs, to much more specific data on populations, road conditions, and weather. The goal is to be better able to predict when and where conflicts will emerge, and to provide accurate information to responders (aid agencies, government officials, etc.) once a conflict erupts. Since receiving the Foundry monies, Carment and his team have expanded the project with funds ($250,000) from the Canadian International Development Agency (Foster 2006). This is one example of how TTOs can help to disseminate knowledge produced at universities that is not strictly commercial in orientation to the general public. Most critically, the Carleton TTO example demonstrates that the social sciences often have very tangible and valuable research to share, knowledge that can help to make our world a better place. It is important to note, however, that compared with other TTO offices, Carleton's resources are paltry. It is also important to note that the Carleton model is most certainly not the TTO model that the majority of Canadian universities are following.

As stressed from the outset of this chapter, universities, like all public sector institutions, are in a competitive environment fighting for public funding. In order for them to continue to attract operational and research money they must convince the public and elected governments of their inherent value to the economy and, more generally, to society as a whole. TTOs can be important institutions for achieving this goal. However, the current configuration of most university TTOs (with emphasis on a more linear model of innovation) and their focus on commercializing patents raises a number of important issues. It is true that the spin-off companies provide an important economic stimulus to the regional economies of the areas in which universities are located, and that the income earned and the number of patents issued is growing. Patents held since 2001 have increased 34 per cent and income earned has increased 6 per cent in the same time (Statistic Canada 2006, 4). However, unless some significant gains are made, these are still relatively insignificant amounts of money.

In addition, one must also factor in the costs of operating the 'standard' type of TTO, the inherent structural problems that Canadian high-tech firms face, and the fact that the time horizons for product development are often between five to ten years. Will universities be willing to wait for a 'payoff' that may or may not ever come? More importantly, will governments (many of whom see commercializing IP as an important output measure) be willing to continue to fund research? It will be interesting to return to this subject in five to ten years to see if, and how, TTOs have adapted to their particular circumstances. This will be especially so if they are unable to increase the income earned from, as well as the amount of, patentable IP. Time, as usual, will tell.

Conclusions

This chapter has examined how Canadian universities are approaching issues of intellectual property rights and developing technology transfer offices as features of managing their new roles in the 'knowledge-based' economy. As indicated, the federal influence via intellectual property law is direct, whereas TTOs emanate from universities' own judgment about how to respond to the varied pressures to disseminate the research and knowledge they produce. The central argument advanced is that intellectual property (IP) generated by universities, particularly patents, and the role of TTOs in disseminating knowledge are two key factors that can not only help to improve the effectiveness and public image of these institutions in knowledge transfer and innovation, but also contribute to their aggregate amount of political capital. In short, they need to maintain overall political and funding support, and political legitimacy in a world where governments face any number of competing demands for resources and political and social attention.

We have seen that TTOs fulfil this role by helping to disseminate university-generated ideas to the external world and that they have also become an integral component of university structure because they help both to commercialize university-generated knowledge in the marketplace and to publicize such commercialization as a tangible measurement of universities' outputs.

The chapter has shown that, while IP in the form of patents has become a greater focus for universities and their TTOs, the earnings from these patents/licensing arrangements have been small even when measured as gross revenues, let alone when net revenues are considered after deducting the costs of managing the IP. It has also shown the variety of IP ownership approaches taken by Canadian universities and the diverse ways in which property rights are defined between the institutions and their faculty members.

The chapter's focus on political capital in relation to IP and TTOs is both a current and a historical argument. As previously noted, the great expansion of Canadian universities in the late 1950s and 1960s was undertaken to ensure that Canada had a competitive, skilled workforce, capable of guiding it into the post-war, industrial phase of its development (Axelrod 1982). While providing affordable, universal access was one result of government investment and university expansion, it was not necessarily the most important factor in the minds of the decision-makers within the federal and provincial state when they decided to fund such expansion. Debate surrounding the function and autonomy of the university, not surprisingly, was often centred on funding levels, and on making certain that universities operated at an arm's-length distance from governments, thus ensuring that the latter did not impose too many stringent conditions on the university's day-to-day operations (Davis 1966, 27–30). These are some of the concerns of those opposed to IP use and more generally to the 'entrepreneurial' university today.

Canadian universities were, however, able to maintain their independence in earlier historical periods. Accordingly, it is likely that they will be able to maintain their autonomy over the coming era as well, even while continuing to fulfil the requirements of the elected governments which provide the majority of their funding. They must be able to strike a balance between their independence and the necessity of adapting to the changing political context, and the needs of the public.

The economic significance of the role universities play in the new knowledge-based economy is stronger now than ever before. Canada must strive to improve the overall quality and quantity of its post-secondary system and the R&D it produces, or risk being surpassed by other developed economies (Canadian Council on Learning 2006). The pivotal role of the universities for ensuring the success of Canada's contemporary economy is undisputed. It is interesting that a state-owned and -operated set of monopolies should play such an important role in the economic well-being of our country, particularly in an era when state-owned institutions are often criticized for being ineffective and inefficient. In some respects, the seminal role of the state-owned universities is harking back to a past era, in which government-owned and -operated institutions played a seminal role in the economic development of Canada (Hardin 1974). It may not be Crown corporations that are leading Canada's economic development today, but the pivotal role of the universities (and thus the state) in such development is apparent. University development is thus consistent with Canada's long history of state-assisted development with respect to economy and industry.

Finally, it is essential to bear in mind that many of the decisions to increase funding to Canadian universities have been made within the upper echelons of both the federal and the provincial governments. Over the last twenty years, an ever-increasing number of critical decisions – including important policy shifts and, of course, decisions that require significant financial resources – are made within central government ministries, outside both the public sphere and the legislature. These include the Premier's and Prime Minister's Office (PMO), respective federal and provincial Ministries of Finance, and the Treasury Board Secretariat (or provincial equivalent). Partisan advisors in Ministers' offices have also had a growing influence, commensurate with their growing influence over the civil service, over such decisions (Savoie 2003).[4]

Remember, too, the opportunity cost of post-secondary funding. One dollar spent on universities is one dollar *not* spent on health care or other policy and public service fields (Laidler 2002, 5). Health care affects all Canadians (especially aging baby-boomers), and is highly valued by Canadians, since universal health care comprises a piece of our national identity. Accordingly, universities are forced to compete for these dollars, despite the fact that they cater to a relatively small segment of the population, when compared to Canada's health care system. Imagine, for illustrative purposes, that decision-makers must choose between building new assisted-living quarters to house people's aging parents, and a new university building. The value of the former, in political terms, could be significantly more than the latter in the eyes of most people and their elected representatives, since care of the aged is (and will become) a pressing concern. This might not be true, however, if university funding were viewed as a seminal contribution to the continued *economic* growth, and the overall material well-being, of our society. Such arguments resonate much better with those charged with making critical public policy decisions and allocating resources. Universities therefore cannot forget the needs of their elected representatives.

---

4 The power of partisan advisors who surround Ministers, Premiers, and the Prime Minister has increased substantially in the last twenty years. Often times they have displaced tenured civil servants in advising the executive of the government on policy. Generally, such advisors tend to be drawn from the government's political party, and as such their loyalties (and appointed positions) are due to their relationship with the party elite. Their interests lie with the fate of their political party, rather than to their particular bureau as with permanent civil servants. Not surprisingly, they tend to bring a short-term 'politicized' perspective to policy creation rather than the longer-term, holistic approach generally brought by career civil servants. The role that Eddie Goldenberg played in Jean Chrétien's government is one example of the increase in power of the partisan advisor.

REFERENCES

AUCC. (2005). *Momentum: The 2005 Report on University Research and Knowledge Transfer*. Association of Universities and Colleges of Canada. Available from: http://www.aucc.ca/momentum/en/_pdf/momentum_report.pdf (accessed 11 January 2009).

Axelrod, Paul. (1982). *Scholars and Dollars: Politics, Economics, and the Universities of Ontario, 1945–1980*. Toronto: University of Toronto Press.

Bird, Malcolm G. (2006). 'Harmful Distraction: The Commercialization of Knowledge at Canada's Public Universities.' In *Innovation, Science, Environment: Canadian Policies and Performance, 2007–2008*, ed. G. Bruce Doern, 281–300. Montreal: McGill-Queen's University Press.

Bramwell, Allison, and David A. Wolfe. (2005). 'Universities and Regional Economic Development: The Entrepreneurial University of Waterloo.' Paper presented at Canadian Political Science Association, Annual Conference, The University of Western Ontario, London, Ontario, 2–4 June.

Brzustowski, T.A. (2006). 'Innovation = Invention + Commercialization: A Systems Perspective.' *Optimum Online* 36 (3) (September). Available from: http://optimumonline.ca/print.phtml?id=261 (accessed 22 November 2008).

Canadian Council on Learning. (2006). 'Canadian Post-secondary Education: A Positive Record – An Uncertain Future.' Canadian Council on Learning. Available from: http://www.ccl-cca.ca/CCL/Reports/PostSecondaryEducation/Archives2006/ (accessed 22 November 2008).

Challis, John R.G., Jose Sigouin, Judith Chadwick, and Michelle Broderick. (2005). 'The University Research Environment.' In *Taking Public Universities Seriously*, ed. Frank Iacobucci and Carolyn Tuohy, 360–75. Toronto: University of Toronto Press.

Crane, David. (2006). 'We Need to Rebalance our 'Asset mix.' *The Toronto Star*, 3 December, 10.

Davis, William G. (1966). 'Ontario and the Universities.' *Governments and the University*. Toronto: Macmillan, 21–46.

Doern, G. Bruce, and Markus Sharaput. (2000). *Canadian Intellectual Property: The Politics of Innovating Institutions and Interests*. Toronto: University of Toronto Press.

Economist. (2007). 'The Sincerest form of Flattery.' 4 April. Available from: http://www.economist.com/business/displaystory.cfm?story_id=8961838 (accessed 4 November 2007).

Etzkowitz, Henry, Andrew Webster, Christine Gebhardt, Branca Regina, and Cantisano Terra. (2000). 'The Future of the University and the University of the Future: Evolution of Ivory Tower to Entrepreneurial Paradigm.' *Research Policy* 29: 313–30.

Fischmann, Brett, M. (2005). 'Commercializing University Research Systems in Economic Perspective: A View from the Demand Side.' Paper presented at the *Colloquium on Entrepreneurship Education and Technology Transfer*. Available from: http://www.entrepreneurship.arizona.edu/docs/conferences/2005/colloquium/B_Frischmann.pdf (accessed 22 November 2008).

Fishman, Ted C. (2006). *China Inc.: How the Rise of the Next Superpower Challenges America and the World*. New York: Scribner.

Foster, Scott. (2006). 'Answering the Call of Millions.' *Carleton Now*, Carleton University, 9 January. Available at: http://www.now.carleton.ca/2006-01/1066.htm (accessed 22 November 2008).

Friedman, Lee S. (2002). *The Microeconomics of Public Policy Analysis*. Princeton, N.J.: Princeton University Press.

Hardin, Herschel. (1974). *A Nation Unaware: The Canadian Economic Culture*. Vancouver: J.J. Douglas.

Iacobucci, Frank, and Carolyn Tuohy. (2005). 'Introduction.' In *Taking Public Universities Seriously*, ed. Frank Iacobucci and Carolyn Tuohy, xi–xix. Toronto: University of Toronto Press.

Laidler, David. (2002). 'Renovating the Ivory Tower: An Introductory Essay.' In *Renovating the Ivory Tower: Canadian Universities and the Knowledge Economy*, ed. Laidler, David, 1–38. Toronto: C.D. Howe Institute.

Leaf, Clifton. (2005). 'The Law of Unintended Consequences.' *Fortune Magazine*, 19 September, 22–7.

Pal, Leslie, A. (2001). *Beyond Policy Analysis: Public Issue Management in Turbulent Times*. Toronto: Nelson Thomson.

Savoie, Donald J. (2003). *Breaking the Bargain: Public Servants, Ministers, and Parliament*. Toronto: University of Toronto Press.

Srikanthan, Thulasi. (2006). 'Ivory Towers No More.' *Toronto Star*, 10 October, 17.

Statistics Canada. (2006). *Statistics Canada Survey of Intellectual Property Commercialization in Higher Education Sector*. Ottawa: Statistics Canada.

Tudiver, Neil. (1999). *Universities for Sale: Resisting Corporate Control over Canadian Higher Education*. Toronto: James Lorimer.

Turk, J.L., ed. (2000). *The Corporate Campus: Commercialization and the Dangers to Canada's Colleges and Universities*. Toronto: James Lorimer.

University of British Columbia. (2003). *Annual Report*, University of British Columbia, Vancouver. Available from: http://www.publicaffairs.ubc.ca/annualreports/2003/annrep03.pdf (accessed 22 November 2008).

Wolfe, David A. (2005). 'Innovation and Research Funding: The Role of Government Support.' In *Taking Public Universities Seriously*, ed. Frank Iacobucci and Carolyn Tuohy, 316–40. Toronto: University of Toronto Press.

8 Federal Government–University Collaboration in the Conduct of Research: Trust, Time, and Outcomes

RUSSELL LAPOINTE

In addition to federal government–university relations that operate through the granting agencies, the CFI's infrastructure funding, and pressures to commercialize and transfer technology, such relations also occur through collaboration in the actual conduct of research. Typically, this involves actions and funding by the federal science-based departments, three examples of which are examined in this chapter. More particularly, the purpose of this chapter is to develop a conceptual framework that helps us to understand more completely the value and limits of partnerships between the federal government and Canadian universities in the conduct of research both in terms of the university's need for independence and the government's needs for S&T that contributes in an effective and timely way to national priorities and policy and regulatory needs.[1]

The framework advanced is centred on three pillars – trust, time, and outcomes, and on two levels of interaction – the interpersonal and institutional levels. The chapter moves analytically back and forth between both the language of barriers and the value and potential for research collaboration across complex institutional boundaries. This micro-level framework is also developed from three case studies of federal–university research collaboration.

In addition to the primary sources cited, the chapter is based on several interviews with scientists and science managers in the three case studies selected for investigation. These were conducted on a not-for-attribution basis. The interviews and the case studies as a whole

---

[1] There are many types of collaborations between institutions. For a good typology of the different types, see Katz and Martin 1997.

add an important practitioner sense of what makes collaboration work or not work. This helps build the micro-level typology employed. The interviews were with university professors and governmental science managers and scientists. It is these individuals who are most involved at the micro level in the creation of research partnerships. It is these researchers who identify, not only potential research projects, but also who they could partner with when developing a research project.

To establish first the terminology that will be used, a 'university researcher' or 'scientist' is generally identified as a professor with a PhD who is teaching and doing research at a university. The 'government researcher' will include not only government scientists but also managers of science and research in particular departments or government laboratories. The 'government manager' may also be a scientist who has moved into the administrative aspect of research. This community comprises those who have hands-on relationships to the research that is being done.

Three university–government research partnerships or collaborations are looked at briefly and illustratively. The first partnership is the Polaris HIV Seroconversion Study, which was a partnership between the HIV Social, Behavioural, and Epidemiological Studies Unit, Faculty of Medicine, University of Toronto, and the Viral Evolution and Molecular Epidemiology Unit of the National HIV and Retrovirology Laboratories (NHRL) of Health Canada. The second partnership is an ongoing one developed between the Communications Research Centre Canada's (CRC) Terrestrial Wireless section and Carleton University's Department of Electronics. The third partnership is between the University of British Columbia (UBC) and the Department of Fisheries and Oceans (DFO) in the research being done at the Centre for Aquaculture and Environmental Research (CAER). This is an extensive collaborative research partnership between UBC and DFO, which is more than a single research project, but instead involves the running and financing of an already existing research centre.

These three research collaborations were chosen because they represented a range of research topics and types of S&T as well as different types of collaborations. The Polaris study was chosen for two reasons. First, it was a significantly large project that was conducted over many years relative to the other two projects. Second, the project was multidisciplinary in nature and dealt with social science issues, whereas the other two did not. The collaboration between CRC's Terrestrial Wireless section and Carleton University's Department of Electronics was chosen because it represented collaborations that were not very long in duration and dealt with a specific scientific area of concern. Also, there was a significant historical relationship between CRC and Carleton University's Department of Electronics. The collaboration between UBC and the DFO expressed in the CAER was chosen because it represented a collaboration that has not set an end date and also involves a significant infrastructure agreement between the two parties. Each of the case studies has a number of unique aspects to it that allows for a survey of barriers that had to be overcome in order for the collaboration to take place.

As will be demonstrated, at the micro level, interpersonal relations of collaboration between academic and government officials are built on trust, but also on other aspects of exchange, pressure, and reciprocity. The establishment of trust in this partnership is based on making sure that there are no barriers to the initiation of this process and no threats to it once it has been established, and that if threats arise they will be dealt with. In short, this chapter is an attempt to draw out and better understand the value, limits, and barriers to fostering this type of collaborative relationship between the federal government and university researchers.

The chapter is divided into two sections. The first section provides a brief preview of the core features of the three case studies of university–federal government research collaborations. The second sets out the main features of the micro-level framework, interspersed with analytical comments about barriers and opportunities derived from the case studies and the interviews. Conclusions then follow.

The Three Case Study Collaborations in Brief

Polaris HIV Seroconversion Study

The Polaris HIV Seroconversion Study was a study that ran from 1998 to 2006 and examined the impact of being affected by HIV (human immunodeficiency virus) in Ontario. The Polaris Study provided a multidisciplinary approach to understanding why some men who have sex with other men get infected and some do not. The goal was to provide an analysis that would lead 'to improved HIV prevention, education testing, counselling, social services and medical treatment' (Polaris 2000). This study was developed out of an extensive survey of AIDS organizations, public health units, and physicians. The goal of the research team was to develop a research approach that would provide 'a more holistic approach to the understanding of HIV transmission' (Polaris 2000).

The Polaris Study, through its multidisciplinary approach, had three main areas of research into HIV and those who live with HIV. First, they studied behavioural and social science aspects of why individuals become infected with HIV and how to live with the disease. They examined how HIV affects different parts of the population from 'women, immigrants, the urban-core disadvantaged, men who have sex with men, aboriginal communities, and injecting drug users.' Second, they focused on epidemiology and examined the sources of new HIV infections and the characteristics of those being infected. This meant that they examined new forms of HIV around the world and viral strains that were resistant to anti-HIV drugs. The third area investigated was that of clinical science. This was the examination of the timing of when to use antiretroviral treatment, in order to help physicians and patients better decide forms of treatment and therapy (Polaris 2000).

CRC and the Advanced Antenna Technology

Research done within the Advanced Antenna Technology section of CRC's Terrestrial Wireless research program concentrates strictly on contemporary antennas. The Terrestrial Wireless research section's goals are to provide an understanding of, create, and produce technologies for wireless systems such as future broadband multimedia wireless systems, military and civilian high data rate mobile wireless systems, personal communications systems, secure voice systems, and integrated electronics (Communications Research Centre Canada 2005a). The clients for this research range from the military and government to wireless service providers, to industry in general. CRC collaborates with many national and international research organizations, including universities.

The Advanced Antenna Technology area of research and development is focused on 'hardware and software investigations pertinent to state-of-the-art, low profile, active and passive antennas and array technologies for applications from L-Band to the millimetrewave band' (Communications Research Centre Canada 2005b). The goals of the program are to develop technology to transfer to industry through the participation of CRC's own researchers, universities and industry. The lab presently consists of twelve engineers, with four of these being adjunct professors.² Seventy per cent of research carried out by CRC is done in conjunction with universities and 30 per cent with private industry.

There are many programs administered through CRC and most research is conducted through a partnership with university professors who apply to get third party funding, usually NSERC funding.³ CRC provides the research facilities, and hardware and software for a particular research project that both the university professor and the program manager agree on. The Advanced Antenna program provides a sophisticated test facility that gives researchers access to 'state-of-the-art antenna measurement and prototyping,' as well as 'an anechoic far-field chamber, a planar near-field scanner, latest commercial and in-house 3D EM analysis tools, and in-house capability for microelectronics fabrication and electronics integration into the antenna' (Communications Research Centre Canada 2005b). In most cases the professor brings in a graduate student, who may be at either the doctoral or master level, to help with the research. The work the student participates in will most likely be related to his or her thesis project. From 1975 to 2006 ninety-eight graduate students and nine universities have worked with the Advanced Antenna Program. The universities that have worked with the program have been dominated by Carleton University, the University of Ottawa, the University of Manitoba, and the Royal Military College.⁴ It is important to note a number of the students have gone to become scientists at the CRC. At the program the idea of mentorship is taken to be an important part of the research relationship.⁵

Centre for Aquaculture and Environmental Research (CAER)

The Centre for Aquaculture and Environmental Research (CAER) is a joint venture between the University of British Columbia (UBC) and the Department of Fisheries (DFO), made formal with the signing of a partnership agreement in 2005. CAER was officially opened as a research centre in 1987. It is a six-hectare site located in Burrard Inlet near the city of Vancouver and a one-hour drive from the University of British Columbia campus. Because of the uniqueness of its location

---

2 Interview with government manager/scientist.
3 Ibid.
4 The other universities are Waterloo, Queen's, Laval, Western Ontario, and McMaster.
5 Interview with government manager/scientist.

CAER has the ability to do both fresh water and seawater research. The Centre concentrates on the general areas of research of marine ecosystems, aquaculture, and salmon and freshwater ecosystems (Fisheries and Oceans Canada 2006).

It should be noted, however, that the research centre had a history of producing research through partnerships before this agreement with UBC. Like many government research facilities, the research centre has gone through many changes in its history. As the federal government proceeded with the Core Review process in the mid-1990s the research centre was scheduled for retirement.[6] It languished for five to seven years after that, with research being done only sporadically. However, with the rise of the aquaculture industry in the late 1990s and early 2000s, the DFO made it a centre of aquaculture research. A significant event that helped to provide motivation for the agreement came in the form of two Research Chairs moving their research labs from the university to the research laboratories at CAER.[7]

Many small collaborations over the years between UBC and DFO led to the creation of CAER in 2005. In the partnership agreement between DFO and UBC the following research areas for CAER were established:

- Regulating fish health protection and the release or transfer of fish into fish habitat or fish rearing facilities;
- Aquaculture enabling technologies;
- Fish health, in general;
- Molecular genetics;
- Cultured/Wild stock interaction;
- Sustainable fisheries and aquaculture;
- Fish nutrition and digestive physiology;
- Healthy and productive aquatic ecosystems;
- Coastal planning and management (Fisheries and Oceans Canada 2006).

This has become a significant research partnership between UBC and DFO. It not only provides research for DFO and the federal government but it is also provides research for the province of British Columbia, which has the responsibility of licensing owners and operators of aquaculture

---

6 Interview with government manager/scientist.
7 Ibid.

farms. The development of a significant aquaculture industry along the coast of B.C. and the subsequent rise of environmentalists opposed to the practices of the industry have pushed both the federal and the provincial governments to determine what are the best practices for the industry in order to limit their impact on the environment.

A Micro-Level Framework for Government–University Research Collaboration

This section sets out the core micro-level framework but also relates it to the case studies, in part because the framework itself, though informed by the literature cited, is based on issues raised in the analysis of the case studies as seen by the researchers interviewed in both the federal departments and the universities involved. Following the introduction of the framework per se, the case studies are linked to a discussion of both the value of collaboration and the barriers to collaboration.

Collaborations are by definition a process of bringing people, groups, or institutions together in order to create or achieve 'something.' That 'something' can be research, an agreement on governance of an environmentally sensitive area, or development of social policy options. What is important about collaborations is that they are a non-traditional way of creating. The functioning of the process involves a relationship between institutions and individuals who in many cases have differing goals but see the possibility of working together as a way to achieve their goals. This means that the functioning of the relationship between partners is extremely important to understand. Collaborations in the pursuit of creating knowledge through research may not have the clear animosity between participants that groups that have opposing or differing agendas, such as a forest company and environmentalists may have, but there still exists a level of apprehension in the relationship that needs to be understood.

In order to examine the relationship between the academic researcher and the government scientist, a framework for describing the structure of the relationship is needed. At the core of the relationship is the underlying difference between the university researcher and the government researcher: the university researcher has been trained to explain, while the government researcher is seeking solutions to problems. There can be overlap, but this fundamental difference makes it critical to understand the relationship between the two. Overall, there seems to be an

Figure 8.1: Government–University Research Collaboration Typology

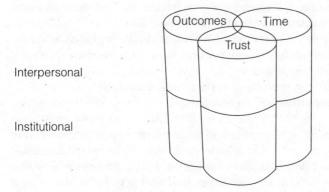

absence of analysis of this relationship at the front-line micro working level in the literature surrounding government and university collaborations. Accordingly, this section sets out a micro-level framework or taxonomy of federal–university collaboration in the conduct of research. There are, of course, some broader macro determinants of the framework; thus, this chapter needs to be seen in relation to chapter 1's discussion of the general features of and tensions in the changing nature of federal government–university relations.

Figure 8.1 illustrates the three pillars and two levels of the framework. Both dimensions will be elaborated on below. The three important pillars in research collaborations – trust, time, and substantive outcomes – are explored. These three pillars will be defined and explained with regards to how they are integral to collaborative research as well as how they operate in both the interpersonal and institutional dimensions of collaboration.

The Three Pillars of Research Collaboration

TRUST
Identifying both the value of, and the barriers to, productive research partnerships between the federal government and a university is a difficult process without understanding and establishing what a good research partnership entails. Throughout the interviews and the literature referred to it is clear that trust is central, in part because where networks

are involved trust is a dominant norm. Trust is an important concept to understand because it is essential to the parties in the collaboration at both the institutional and interpersonal dimension.

In a simplistic sense of the word trust is usually conceived as an expectation that someone will take action to fulfil a promise that they have made. Keith J. Blois argues that trust is much more than this (Blois 1999). He argues that the simple expectation of the fulfilling of a promise is not trust but rather a form of reliance. He argues that 'what distinguishes trust ... is the expectation that the other party may take initiatives (or exercise discretion) to utilize new opportunities to [the other's] advantage, over and above what was explicitly or implicitly promised' (Blois 1999, 198). Blois points out that trust has an 'emotive element' for the individuals involved, and that if trust is broken by a failure to meet expectations then there exists a sense of betrayal. He further points out that trust is based on a 'stated commitment' that is about the other's dependable good will to go beyond one's contractual obligations to produce even more than what has been agreed upon. Thus, what is important is what the individuals feel about each other regarding this larger sense of trust.

Regarding institutions, Blois poses a significant question, namely, can organizations as a whole 'trust'? He points out that most literature on this subject speaks of organizational trust as a given. In short, if individuals trust, then organizations trust. Blois argues that what is important in this issue is the role that individuals play in their organization, because their behaviour might be different if they were not acting in the role they have with the organization and instead were acting as individuals. Also, the expectations of the other within the role one is fulfilling significantly influences trust, especially when meeting for the first time. However, Blois concludes ultimately that it is the individual who 'trusts,' not the organization. Nonetheless, this 'points to a need to ensure that all those in an organization whose actions might affect its reputation' are aware of what actions are appropriate to maintain the reputation of the organization (Blois 1999, 200). Trust of one organization by another is thus also about the reputation of the other organization, but with reputation built upon the behaviours and actions of individuals. This leads to a need to examine the interpersonal level.

Trust is a product of how researchers get along at an interpersonal level and their actions defining the reputation of the institution to which they belong. The interviews provided evidence that this was the case with the three collaborations looked at. Even in the larger research partnerships, like the one with CAER, a sense of trust had to be established in order for that agreement, as well as the roles that important individuals played, to work. For example, the individual actions of the two UBC professors with Canada Research Chairs moving their offices and research to the facility played an important role in demonstrating not only their own commitment, but also that of their department and ultimately of the university as a whole to the collaboration. While trust might be thought of, as one interviewee put it, as being a matter of 'I will scratch your back and you scratch mine,' it is clearly more than that.[8] Trust at the interpersonal level is what can be looked upon to deal with barriers to the development of collaborations. Trust is what sustains the life of a collaboration in times of trouble.

Barriers to the creation of trust can be any number of things that affect the perception of the relationship by the other partner. For example, actions such as taking too long in responding to a request, the frequent movement of key personnel, and changes of leadership both at the political as well as the administrative level can create a sense of uncertainty that hampers the creation and the sustaining of trust.

Trust interacts with the other two pillars – time and outcomes – which also affect the building and sustaining of a core sense of trust. The interaction among the pillars happens both at the interpersonal and institutional levels.

TIME

Time plays an important role in the creation of collaborations as well as in evaluating the success of collaborations. When discussing time it is best to discuss it almost immediately in terms of how it interacts with the other pillars. Also, it is important to consider time as a resource, and one that is unique and valuable to those involved. Time and trust interact in two ways: first, the time it takes to meet others to do collaborative research; and second, the time it takes to create the trust needed to build the collaboration.

Collaborations involve not only the technical skills that each person possesses, but also the human resource skills of working with others that may take time to learn, depending on the level and the extent of the experience of the participants with collaborations (Katz and Martin 1997, 14). This is not to say that all collaborations are the same, but rather that there will be less to learn and more confidence in handling difficult

---

8 Interview with government manager/scientist.

situations the more the researchers have participated in collaborations. As Katz and Martin stress, the process of collaborating is a learning by doing process that can save researchers time in advancing their own knowledge of an area and enhance their skills as researchers:

> It can be time-consuming for an individual to update their knowledge or to retrain. Furthermore, not all the details concerning new advances are necessarily documented. Much of the knowledge may be tacit and remains so until researchers have had the time to deliberate and set out their findings in a publication. Frequently, considerable time elapses before the knowledge appears in written form. Collaboration is one way of transferring new knowledge, especially tacit knowledge. Furthermore, research requires not only scientific and technical expertise, but also the social and management skills needed to work as part of a team. These cannot be readily taught in the classroom – they are best learned 'on the job' by engaging graduate students or young postdoctoral researchers in collaborative activities. (Katz and Martin 1997, 14)

The amount of time that is taken in engaging in the many aspects of collaborations will be judged by what value the process brought to the creation of new knowledge for the researchers.

There are many ways in which individuals and institutions can come in contact with each other in order to create research collaborations. However, much depends as well upon the initial impetus to do the research. If the research is being driven by the government through decisions mapped out in throne speeches, budgets, or policy decisions, then more than likely there will be a faster pace to finding partners with whom to collaborate. The top-down approach also means that barriers to this part of the process would most likely be dealt with faster.

The counterpart to the top-down movement of research would be the bottom-up research partnership that begins with ideas that commence with government scientists or university researchers discovering each other and developing an idea for a research project. This step in the overall process can occur in many arenas: conferences, literature reviews, or word of mouth within a particular science network. Here is where the beginnings of trust building and the foreshadowing of potential outcomes take place. Time is needed to build a personal level of confidence in the other individual's level of knowledge in an area and also in this individual's ability to deliver on what they said they would or could contribute. Here, as mentioned above, the human resource skills of working with others will help. Learning how to deal with other people in a collaborative process can be time consuming, but it can also be fruitful not only for the current project but for future undertakings as well.

Once the identification of a potential partner has occurred this will lead to a need for a dialogue. This dialogue does not have to occur face-to-face given that there are many ways to communicate. However, again this will take some time. The next step would be to approach officials/scientists at the senior levels of departments to see if they would be interested in collaborating on a potential project. The bottom-up approach contains a 'selling' process that can be very time consuming. Also, within this 'selling' process is the convincing of one institution of the other's ability to perform and successfully complete the project. Within this process barriers are going to be harder to deal with because there might not be the impetus to make them more flexible, as would be the case for top-down approaches. The hierarchical structure of power within the organizations can be either a benefit or a hindrance depending upon what path the collaboration takes and how long it requires to evolve and mature.

In both the top-down and bottom-up paths of research creation, collaboration is dependent upon this time-centred process of developing and maintaining trust. As Blois points out, trust is created through experiencing the commitment of the participants to the project as evidenced by actions that can be seen and demonstrated. He continues by noting:

> Trust evolves through the process of a growth of knowledge and understanding of the people with whom we interact plus the actual experience of working with them. Part of getting to know somebody is becoming aware of the extent of their commitment to us plus the circumstances under which we can trust them. (Blois 1999, 213)

This process takes time, energy, and resources. If there are barriers to developing trust then it will be a longer process. This will also lead to an evaluation of whether the effort and time put into the collaboration was worth what was eventually achieved in the outcomes of the research.

In general, time is needed to start, develop, maintain, and conclude collaborations. Katz and Martin created a brief list of these and other possible issues, but emphasized that the point that maybe the most important resource in the collaboration process is time (Katz and Martin

1997, 15). Thus, the use of time as a resource becomes a factor in the evaluation of the outcomes from the collaboration, as will be seen in the interaction of time and outcomes discussed below.

OUTCOMES

In this typology, outcomes are the concrete products of the collaboration. They can be categorized in two ways. First is the actual research produced in the form of new knowledge or understanding of a specific area. More specifically, these are products such as policy options or solutions that can be applied to problems facing government, expressed in peer-reviewed journal papers, chapters in books or an entire book, and any tactical evidence of the collaboration.

The second category is outcomes that provide secondary benefits that are not tied directly to the knowledge created through the research but are in some cases just as important (Katz and Martin 1997, 15) On the interpersonal level is the feeling of accomplishment as well as a form of recognition of what was produced by superiors or peers. Also, with this acknowledgement there might be financial or career benefits. For the university researcher, the creation of this knowledge will help in their pursuit of tenure or a higher status within their department or faculty. For the government researcher, as mentioned above, secondary benefits are less obvious. However, it is highly likely that superiors would recognize this work in some way in evaluating the potential of an employee. The reverse is also true if the project is a failure and outcomes do not measure up to the resources and effort put into the project. This can damage a person's reputation as a researcher and reduce the likelihood of their participation in further collaborations.

The institutions benefit from the production of knowledge as well, and successful outcomes arise not only from the use made of the new knowledge but also from the enhancement of their reputation. The creation of a good reputation allows the institution to 'sell' its services and expertise either internally or outwards to others seeking to do collaborative work. For the university this can mean benefits such as attracting leading-edge faculty and graduate students and the possibility of further research funding from both the public and private sector.

At the interpersonal level, what is important to the successful outcomes of the collaboration is how the participants feel about it. This evaluation can be seen in the willingness of those involved to work together again. From the interviews it emerges that a formal process to evaluate collaborative outcomes often does not exist, but there always exists an informal one. The informal evaluation process consists of the researchers and senior officials asking themselves whether the outcomes and benefits of the research are commensurate with the amount of resources put into the collaboration.

Outcomes and the collaborative process are evaluated in terms of factors of both trust and time, and of course on the substantive aspects of the science and technology involved. Trust is involved because after an initial experience trust would either be created or not. The existence of trust would most likely lead to the undertaking of other projects because expectations for the current or recent one had either been met or were being met or exceeded. Time, as a resource, is involved in a cost-benefit analysis simply by asking the question whether the amount of time, effort, and resources put into the project have been worth what the substantive outcome has given to both the individual and the institution. This would be of particular importance to bottom-up research projects because senior decision-makers might be reluctant to assume responsibility for the project. Even though the research process may have been completed successfully with the production of new knowledge, the evaluation of the process is what will determine whether the participants will embark on further collaborative ventures in the future. Therefore, how barriers were dealt with and the amount of resources required to overcome them will be essential to viewing the collaboration as a success and the possibility of new research being undertaken. As one academic researcher said of a previous collaborative research project in which she had participated, 'It successfully created knowledge but it was simply too difficult in getting it started to do it again.'[9]

Those interviewed for this chapter all stated in one way or another that trust is the cornerstone of a collaborative relationship. Trust is expressed through actions such as how they personally got along, whether deliverables were met from both sides, and whether each side had a clear understanding of each other's roles and objectives in the research project. Thus, though trust is a favoured starting point for discussing successful collaborations, the criteria quickly move on to issues of time and substantive outcomes.

Interpersonal and Institutional Dimensions

As described above, the three pillars of collaboration interact at both an interpersonal level and an institutional level. The interpersonal dimension of the relationship is the one that the researchers have with

---

9 Interview with university researcher.

one another. It can simply be, as one university researcher pointed out, the need to work with someone 'nice' or 'interesting'. However, as described above, it is the creation of a bond of expectations and commitment that ultimately matters.[10] The creation and execution of a collaboration is a process involving several steps and factors that begins with meeting other individuals who share an interest in similar research areas. Katz and Martin point out that there are endless reasons for collaborations and this is because 'collaboration is an intrinsically social process and, as with any form of human interaction, there may be at least as many contributing factors as there are individuals involved' (Katz and Martin 1997, 4).

Regardless of whether the research is applied or theoretical in nature, getting research partnerships started begins with ideas and the discussions of those ideas between individuals who share a common interest and/or goal. Partnerships arise through networking situations such as conferences, published and unpublished literature, and word of mouth between colleagues. Although the internet and electronic sources have allowed people more access to information, the ability to exchange ideas leading to the creation of research collaborations is frequently dependent upon the creation of some form of substantive personal interaction. This is also because tacit knowledge acquired through interpersonal understanding and observation of how a person works and communicates knowledge is central to the process. Knowing the person one is going to work with on at least a professional level is important for the creation of trust. As the federal Council of Science and Technology Advisors (CSTA) reported in their publication *Linkages in the National Knowledge System: Fostering a Linked Federal S&T Enterprise (LINKS)*:

> Regardless of the institutional and management structures in place, the success or failure of collaborative S&T initiatives often comes down to the human factor. Fostering personal relationships and information and knowledge sharing among partners is fundamental to success. It helps to build trust and 'buy in' among participants and to foster a shared culture and common lexicon. (Council of Science and Technology Advisors 2005, 7)

The creation of trust will occur more often than not in informal ways such as attendance at conferences, email or phone exchanges, and references from others. As a government official stated, their field of advanced

---

10 Interview with university researcher.

antennae research in Canada is limited to about two hundred experts, almost exclusively academic and government people, and this made their biannual conference extremely important.[11] As Katz and Martin describe, it is at these places that those who share the same social level in their fields meet, and this decreases the spatial distance between potential collaborators (Katz and Martin 1997, 7–8).

From the creation of the relationship to the relationship producing the actual research there is the interpersonal dimension to building and maintaining a level of trust. Learning how to work with others is a process of understanding and experiencing their strengths and weaknesses. This means also that individuals will have to develop the ability to collaborate on research papers, time management skills, communications skills (especially when communicating to individuals in a different institutional setting), and other interpersonal and social skills needed to work successfully with others.

The other dimension is institutional. This dimension was described by the interviewees as being the originator of most of the barriers to collaboration. Katz and Martin confirm this as they state:

> At the most basic level, it is people who collaborate, not institutions. Direct co-operation between two or more researchers is the fundamental unit of collaboration. However, we often talk about collaboration at other levels – between research groups within a department, between departments within the same institution, between institutions, between sectors, and between geographical regions and countries. Indeed most policies are aimed at fostering collaboration at these higher levels rather than inter-individual collaboration. (Katz and Martin 1997, 16)

Organizations tend to push for the use of research collaborations to help them fulfil their mandates in the most cost-effective way. However, sometimes the process of collaboration does not fit well with the stringencies of an institution. Katz and Martin place as much importance on the individual within the institution as Blois does around the issue of trust. Both are demonstrating that ensuring that the interpersonal relationship between the individuals works is the key to the success of collaborations. This makes it of the utmost importance that the institutional dimension function in ways that do not interfere with the interpersonal dimension and hamper the conduct of research.

---

11 Interview with government manager/scientist.

For a barrier to be classified as institutional, it has to have certain qualities. First, it must be a factor in determining how individuals conduct themselves in their roles. The person's institutional role and job description, as well as their position within the hierarchy of power and decision-making, affects the responsibilities and behaviour they bring to the collaborative relationship. This is something that may have the most impact on those from government, but it applies to academic researchers as well, as universities also have policies restricting what kinds of research they will allow and research ethic boards at universities influence the role of the academic researcher.

As noted earlier, the mandates and policies of the government departments and labs present the most prominent potential barriers to research collaboration (Doern and Kinder 2007). Such mandates may well directly conflict with the researcher's academic freedom and also with potential intellectual property rights.

The path of research creation, whether it reflects a top-down or bottom-up approach, is most visible in the institutional dimension. In the path of research development the two dimensions interact, particularly in bottom-up research. Having the ability to create ideas and research projects from the ground up is important because this is the type of research that researchers often become most connected to and feel most motivated to conduct. Because of this there is a personal stake in this type of research development. This is not to say that there will not be a personal connection to top-down research, but the personal ownership of that research is more evident in a bottom-up approach.

The other prominent aspect of the institutional dimension is the process of accountability, most significantly in the area of demonstrating how public funds were used within the research project. More on this issue will be explained below, but the prominence of the Office of Auditor General and new accountability policies has meant more time and resources have to be dedicated by all participants to this part of the research process.

Barriers and 'Beyond Barriers' in Understanding Research Collaboration

There is a need to discuss key issues about the positive value and nature of research collaboration. The benefits of collaboration are many, both to the individuals and the institutions, but as Katz and Martin point out there are costs, as already indicated, of increased administration, individual and institutional learning, and specific issues of mandate and policy compliance as seen in matters such as intellectual property rights. These costs can be equivalent in their effects, and each can have a point of origin with one of the participants. Because of this there are differences in the ability to identify and manage this type of barrier, and hence to potentially convert it into an overall opportunity for collaboration. For this reason it is particularly important conceptually to differentiate internal barriers from external barriers.

Internal barriers derive from aspects of the relationship that the government scientist or manager or the university researcher has with their *own* institution that affect the collaborative effort. In the case of the government scientist or manager, internal barriers consist of procedures such as acting in compliance with accountability legislation if the use of government funds or resources is involved. Other issues such as research ethics boards and the like can also provide a barrier to the creation of research collaborations. The hierarchical relationships within the institutions is a significant influence here. Pressures from the hierarchical authoritative structure of an organization influences the practices and procedures by means of which the organization will work with others.

The university researcher faces some of the same type of institutional barriers, such as formalized procedures for the use of granting council funding or other third party funding that is administered by their university. The use of graduate and undergraduate students as research assistants also creates other complications, in that, while a research project can help the university professor in their role as a dissertation supervisor by providing a project out of which the student can produce a dissertation, there still exists a financial and administrative responsibility as well.[12]

Also, within both types of institutions internal funding and costs are an issue. Having up-to-date research facilities means having the capital to maintain them in this condition, and this is an issue for both university and government researchers. Additionally, both government scientists or managers and university researchers have been affected by the overall changes to how the federal government funds research in the country. As previous chapters have demonstrated, the innovation agenda of the Chrétien and Martin Liberal governments as well as the current Conservative government has meant different things to different

---

12 Interviews with university professors and government scientists and managers.

parts of the research community in Canada. For example, this innovation agenda has meant an increased emphasis on the commercialization of research (Polster 2004).

External barriers are things that affect the collaborative relationship but are located with the other partner in the relationship. Or, they can include barriers that might be thrown up by third parties outside the relationship, such as other government departments or other research institutions. For the purposes of this chapter, external barriers can be categorized into four types: the first deals with the mandates of the two partners; the second with the issue of timing; the third with the specific issue of Intellectual Property Rights (IPR); and, finally, the fourth barrier deals with outcomes and the dissemination of research findings.

Mandates, specifically government/departmental mandates, can create a barrier if there is a conflict between what the department's function is and the university researcher's need for academic freedom. The university researcher conducts research with the notion of having academic freedom and independence that allows for an objective examination and report of findings. This may potentially conflict, not only with departmental policy, but also with overall government policy. In order for a collaboration to be created an understanding of the purposes of the two partners regarding the research must be made clear. As the Council of Science and Technology Advisors (CSTA) has argued, the selection of research projects within industry, universities, and government are based on different criteria:

> In industry, research projects are selected based on their anticipated contribution to new products, processes, and services. Universities and Canada's granting councils use peer review to make competitive project selection decision based on scientific merit. In addition to scientific merit government projects should demonstrate independence, alignment with government and departmental mandates and stakeholders needs, transparency, openness and ethics. (Council of Science and Technology Advisors 2005, 6)

Understanding what the other partner needs out of the research partnership and the limitations that they bring into the relationship will bring about a more productive outcome.

What is also important is the hierarchical path that the research is taking. If this is a top-down research path where senior authorities or political leaders are asking for this research to be done this top-down pressure on the department creates an incentive for the department to work hard to navigate the research process around potential barriers.

Also, with differences in mandates there exist differences in the time frames that research is done in. Times frames are identified as the different time pressures that each researcher faces, whether this be the yearly cycle of both institutions, such as the annual budget cycle that government imposes on departments, or the way a university divides up its academic year, or the immediate action that a political event might impose. These differences can become a barrier if the additional time-centred pressures that they place on a partner are not recognized.

One other external barrier includes dealing with the use of graduate students to conduct research on a government site. Some of the issues concerning students working on site may include criminal record checks, worker compensation issues, and the use of international students in a project. This does not mean that international students are excluded from doing research on collaborative research projects, because some research can be done at the university. Also, unlike the situation of the university professor mentioned above, the role of the government scientist in mentoring or working with students on their dissertations is not acknowledged or accredited in any official way.

The ability to locate the suitable partner for a research collaboration is also a potential barrier, as the process of bringing together people who can be of mutual assistance to each other is one that can be very difficult. This process of research and policy incubation happens, as we have seen, through conferences, literature, various forms of media communication, and also by word of mouth.

Conclusions

This chapter has explored key relationships between the federal government and universities that are established through the actual conduct of collaborative research. Three university–government research partnerships have been looked at briefly an illustratively. The chapter has developed a conceptual framework that can help understand more completely the value and limitations of partnerships between the federal government and Canadian universities in the conduct of research, both in terms of the universities' need for independence and the government's needs for S&T that contributes in an effective and timely way to national priorities and conforms with policy and regulatory needs.

The micro-level conceptual framework advanced in the chapter has arisen inductively from the three case study illustrations and interviews, but also deductively from some of the literature cited and from some of the inherent macro issues discussed in chapter 1 and other earlier chapters in the book. The framework developed suggests that each of the three pillars – trust, time, and outcomes – are important and interactive in nature. And they must be related to both the individual and interpersonal levels of interaction and collaboration.

This suggests, not surprisingly, that university–federal government research collaborations involve both complex and dynamic relationships. The horizontal dynamic of cooperation between institutions is at the same time influenced by the hierarchical pressures of authority and accountability. This dynamic creates opportunities as well as barriers. Collaborations rest on the pillars of trust, efficiency of the use of time and resources, and the substantive outcomes of the relationship.

These three pillars also work within the interpersonal dynamic of building trust between people, learning how to work together, and evaluating whether the collaboration has produced a successful outcome. The institutional dimension also works with the dynamic of building trust between institutions, learning to work with another institution, and determining whether the effort and time of the organizations is being used wisely. However, it is necessary to reiterate the crucial importance of the interpersonal dynamic among the individuals working within the institutions that are participating in collaborations.

The barriers that have been discussed above are all manageable to some degree, but they remain significant features that take on different shapes and rankings between collaborative partners as they negotiate a partnership and carry it out and then later assess it. However, what is essential to building effective and lasting collaborations and ultimately good research is the creation of some kind of basic underlying trust. Trust is the component that will hold the collaboration together through difficult times in the research process, particularly at the beginning. The creation and protection of trust within a collaboration, the amount of time it takes researchers to deal with substantive outcomes, and the issues involved in building and managing research projects are the three main themes that emerge from our examination of the case studies.

What this really means is that each side of the collaboration recognizes the needs of – and the pressures on – the other partner or potential partner. University researchers must recognize the vertical pressures of accountability on their counterparts in the government. Government scientists and managers must in turn recognize the needs of university researchers for academic freedom and the requirement that they publish in order to achieve advancement in their careers.

Overall, this chapter also argues that barriers to research collaboration need to be assessed and understood on the basis of whether they are internal or external, and solutions to them developed accordingly. As has been demonstrated, internal barriers for university researchers are raised by issues such as the overhead costs created for universities as recipients of research grants and by the administrative accountability that is required for research. Internal barriers faced by government scientists/managers include aspects such as departmental mandates, the administrative accountability process that includes audits from other departments, the time frame demanded by the issue, and the internal funding for research and research infrastructure.

The external barriers faced by the university researcher are those of finding the right research partner with whom to create a research project, the mandates of the department they are working for, the organizational culture of the department, the funding of the department for research and research infrastructure, the external accountability process that includes ethics board approval, and IPR. For the governmental scientist/manager the external barriers are similar to those faced by the academic – in the problems that they face in finding partners, in the organizational culture of their department, and also in the issue of IPR. All these barriers have a further distinguishing dimension related to whether they can be handled independently by negotiating with their partner or have to be dealt with internally, in which case the partner has to accept what is happening if it cannot be altered significantly, or at all.

The barriers that have been discussed in the chapter are not, as stated earlier, insurmountable. Every barrier has a solution, although some are more manageable than others, and potential problems need to be kept in mind when collaborations are entered into. Research in the S&T community can only get better with the introduction of new and different ideas and perspectives on issues, and collaborations can create the conditions for this to happen.

The analysis has shown overall that there are many practical puzzles, barriers, and applied values in cross-institutional research collaboration, but the micro-level realities highlighted through the framework and case studies can only be worked through by those persons in government and in universities who have to conduct the research as well as manage all of the trust, time, and outcome-related issues that occur at both the institutional and the interpersonal levels.

REFERENCES

Blois, Keith J. (1999). 'Trust in Business to Business Relationships: An Evaluation of its Status.' *Journal of Management Studies* 36 (2): 197–215.

Communications Research Centre Canada. (2005a). 'Terrestrial Wireless Systems.' Communications Research Centre Canada. Available from: http://www.crc.ca/en/html/crc/home/research/wireless/mission (accessed 5 June 2005).

– (2005b). 'Antenna Test Facility.' Communications Research Centre Canada. Available from: http://www.crc.ca/en/html/crc/home/research/wireless/antenna_facility (accessed 5 June 2005).

Council of Science and Technology Advisors. (2005). *Linkages in the National Knowledge System; Fostering a Linked Federal S&T Enterprise (LINKS)*. Ottawa: Industry Canada.

Doern, G. Bruce, and Jeffrey Kinder. (2007). *Strategic Science in the Public Interest: Canada's Government Laboratories and Science-Based Agencies*. Toronto: University of Toronto Press.

Fisheries and Oceans Canada. (2006). 'Center for Aquaculture and Environmental Research.' Ottawa: Fisheries and Oceans Canada. Available from: http://www-sci.pac.dfo-mpo.gc.ca/sci/facilities/westvan_e.htm (accessed 4 September 2006).

Katz, J. Sylvan, and Ben R. Martin. (1997). 'What is Research Collaboration?' *Research Policy* 26: 1–18.

Polaris HIV Seroconversion Study. (2008). 'About the Study: New Directions in HIV Prevention and Care.' Available from: http://www.hivpolaris.org/ (accessed 10 October 2008).

Polster, Claire. (2003–04). 'Canadian University Research Policy at the Turn of the Century. Continuity and Change in the Social Relations of Academic Research.' *Studies in Political Economy* 71/72 (Autumn–Winter): 178–81.

Smith, David, and J. Sylvan Katz. (2000). 'Collaborative Approaches to Research.' HEFCE Fundamental Review of Research Policy and Funding – Final Report, April 2000. Available from: http://www.sussex.ac.uk/Users/sylvank/pubs/collc.pdf 9.

9 The Co-Location of Public Science: Government Laboratories on University Campuses

JEFFREY S. KINDER[1]

The science policy dialogue in Canada often seems to dwell on the basic rationales for public sector support and performance of research and development (R&D), i.e., 'Why public science?'[2] I wish to move beyond this question about fundamentals by acknowledging that the federal government has a legitimate, indeed critical, role to play, not only in supporting academic research but also in the conduct of science in the public interest (Doern and Kinder 2007). Other questions then present themselves, including what should be the priorities for public science, i.e., 'Which public science?' and what is a sustainable level of support, i.e., 'How much public science?' These questions are beginning to receive more attention in Canada. However, an additional issue that merits more analysis is the appropriate institutional location or 'siting' of public sector research. In other words, in this chapter I focus on the question of 'Where public science?' and in particular on the greater explicit consideration recently being given by the federal government

---

1 I gratefully acknowledge the input provided by Mark Brooks, Tom Brzustowski, Erin Cassidy, Michelle Crawley, Philip Enros, Ron Freedman, Margaret Hill, Ingrid Schenk, Pierre Therrien, and the individuals interviewed for the case studies. The chapter also benefits from the comments of the editors and reviewers. Remaining errors are mine.
2 Clearly, as discussed in chapter 1, the main preoccupation has been on how to increase private sector R&D. For the purposes of this chapter, I label the R&D performed in the federal government and in public institutions of higher education as 'public science' or 'public sector research' to distinguish it from the R&D performed by the private sector, where 'D' (technology development) is much more prevalent than in the public sector. The public/private distinction can be misleading in this era of policy pressure on public science to find alternative revenue sources, including private sector funding, and to increase support for the commercialization of research. However, the distinction remains useful for my purposes here.

to co-locating federal labs on university campuses. This adds a further dimension to the kinds of issues discussed at a more micro level in LaPointe's analysis in chapter 8 on federal–university collaboration in the conduct of research.

Universities are increasingly important players in Canada's national science and innovation system. In 2004, for the first time, universities – not government laboratories – were the primary recipients of federal R&D funding (Statistics Canada 2006a). This is largely the result of the steady increases in federal funding for academic research since 1997, while funding for federal intramural R&D has remained relatively stable. But it can also be seen as a continuation of a decades-long trend in the post-war period in which the traditional dominance of federal research performers has given way to other performers (Kinder 2003).

As Canada's science and innovation system has matured, research capacity has become more distributed, to the point where no institution can 'go it alone.' Embracing 'systems' thinking means recognizing that the benefits of the national science system emerge 'not only from more R&D but from strong partnerships, networks, and linkages' that facilitate information flows, knowledge transfer, research collaboration, and the training and mobility of highly qualified personnel (de la Mothe 2000, 34). In many cases, public science is best advanced, not through the federal intramural laboratories or universities working in isolation, but through linkages among them and with other research performers.

The purpose of this chapter is to begin to explore one form of government–university partnering – the co-location of government laboratories with university campuses. The analysis is timely given the interest in co-location on both sides of the relationship. On the federal side, the Harper government has considered transferring the management of some federal labs to universities (Government of Canada 2007). On the other side, the Association of Universities and Colleges of Canada has surveyed its members to inventory the federal research establishments they host (AUCC 2007). Key questions that this chapter seeks to address include:

- Where is public sector research performed and why?
- What are the theoretical/historical reasons for siting public research in government labs or in universities?
- What is driving the current trend towards co-location of government laboratories with universities?

- What are the benefits and challenges of co-location from both the host university's and the federal government's perspectives?
- Is co-location a model for the future of public science?

The exploratory nature of this chapter must be emphasized. Its analytical breadth and depth are limited for a number of reasons. First, as part of a broader research program on government laboratories, the chapter represents only an initial look at the practice of co-location. Co-location has not been the focus of much academic attention and the literature is quite thin, consisting mostly of grey literature (e.g., government studies). This chapter seeks to add to the understanding of co-location and its policy implications for federal government–university relations.

Second, the chapter relies on three case studies, which are only briefly profiled here. The level of collaboration existing in federal–university co-locations ranges from the highly integrated joint initiative (e.g., the National Institute for Nanotechnology) to the lone federal researcher located on a campus (e.g., a single scientist of the federal agriculture department's Atlantic Food and Horticulture Research Centre located at the Université de Moncton). Here I have chosen to examine co-locations involving significant federal presence on campus, including a distinct facility. Specifically, I examine the following co-location relationships:

- The *Saskatoon Research Centre (SRC)*, a unit of Agriculture and Agri-Food Canada, which along with its predecessor organizations has been located at the University of Saskatchewan for almost a century.
- The *National Wildlife Research Centre (NWRC)*, a research establishment of Environment Canada that was relocated to the campus of Carleton University in Ottawa in 2002.
- The *National Institute for Nanotechnology (NINT)*, located at the Edmonton campus of the University of Alberta. NINT is a partnership between the National Research Council, the Government of Alberta, and the university that goes beyond mere co-location.

The cases were selected so as to provide a representative sample of research institutions from different federal departments and host universities. The labs perform a variety of scientific functions, ranging from fundamental investigations to mission-oriented applied research to regulatory science. They also vary in their experience with co-location, with the NINT and NWRC being recent co-locations, while the SRC has a long history of co-location. Another difference is that the SRC and the NINT were newly created as co-located organizations, whereas the NWRC previously existed as a stand-alone federal establishment. Other examples of co-located labs could have been chosen, and there are risks in generalizing from these specific cases.

Finally, the research relies primarily on a review of the science policy literature and documents related to the three labs. Initial findings from these sources were tested through a small number of anonymous interviews with key officials currently or formerly associated with these co-location arrangements.

The chapter is organized into four main sections. I begin with a review of the conceptual underpinnings for the siting of public sector research. In the second section I examine the contemporary drivers for co-location. The third section profiles the three case labs. In the fourth section, I analyse the benefits and challenges of co-location from both the host university's and the federal government's perspectives. In conclusion, I consider co-location as a model for the future and suggest future research directions.

Where Public Sector Research?

To begin to answer the question of where public sector research is performed in Canada it is useful to look at the data on national R&D expenditures (see the book's Appendix). The official data indicate that in Canada, the performance of R&D is primarily the province of three sectors: industry, academe, and the federal government.[3] While business enterprises are the largest performer of R&D, with 54 per cent of the gross domestic R&D expenditures, universities and affiliated institutions are also major players, performing about 36 per cent of the national R&D activity. As chapter 1 has shown, the federal government has seen its share of R&D performance decline over the last decade and now performs less than 8 per cent of the national total (Statistics Canada 2006b).

Gaining a deeper understanding of the historical rationales for where, organizationally speaking, public sector research is conducted is aided by first understanding the traditional breakdown of research activity.

---

3 Other R&D performers, including provincial research organizations and private non-profit organizations, can and do make important contributions particularly in niche areas, but the vast majority of national R&D expenditures are performed in the 'big 3' sectors.

Here, as with so many aspects of science policy, a good starting place is the influential report by Vannevar Bush, *Science: The Endless Frontier*.[4] Bush's report to President Truman in the closing days of World War II is generally regarded as the single most influential document shaping U.S. science policy in the second half of the twentieth century. The report's basic tenets, if not its specific policy prescriptions, have also been highly influential in Canadian science policy (Kinder 2003).

The Nature of Scientific Activity

The Bush report divides scientific activity into three broad categories: (1) *pure* or *basic research*, (2) *applied research and development*, and (3) an intermediate category labelled *background research*. It is important to note that the report immediately cautions that the boundaries between these categories are not rigid or sharply defined and that it would often be difficult to locate any particular research project within a single category. 'On the other hand,' the report continues, 'typical instances are easily recognized, and study of them reveals that *each category requires different institutional arrangements for maximum development*' (Bush 1945, 81; emphasis added). Thus, Bush felt it important to understand the different types of research activity and the organizational settings in which each kind flourishes.

What the report refers to as *pure research* or *basic research*, and what has since been referred to variously as *fundamental, curiosity-driven,* or *discovery-oriented* research, is research without specific practical ends in mind. The Bush report argues that the novel nature of pure research 'makes desirable the provision of rather special circumstances for its pursuit' (Bush 1945, 81). In particular, the unpredictable course of pure science, with its multiple pathways to discovery, and the long-standing convention that these discoveries be immediately placed in the public domain all suggest a special setting in which researchers have the freedom of mind to pursue and report new knowledge. Bush famously wrote that scientific progress results from 'the free play of free intellects, working on subjects of their own choice, in the manner dictated by their curiosity for exploration of the unknown' (Bush 1945, 12). The report concludes that pure science 'does not always lend itself to organized efforts and is

---

4 In referring to the 'Bush report' in this section, I am including the supporting report of the (Bowman) Committee on Science and the Public Welfare as well as the summary report written by Bush.

refractory to direction from above' (Bush 1945, 81). Bush felt that universities, with their tradition of academic freedom, provide the ideal environment for such research.

In contrast with basic research, *applied research and development* is more amenable to organized effort. With applied research and development, the specific objectives and milestones can typically be mapped out in advance, and performance can be judged against these. If successful, the outputs of such activity are of definite practical or commercial value; therefore, applied research and development is deemed to be primarily the domain of the private sector. In addition, government agencies are often engaged in applied research in support of their particular missions. In sectors such as agriculture, where the private sector producers tend to be small and widely dispersed, the profit motive may not ensure adequate investment in R&D, and so government has traditionally filled an important role in providing the necessary applied research and development (Bush 1945, 83).

Finally, between these two categories of scientific activity, the Bush report identifies an intermediate category labelled *background research*. This category includes scientific activities that today are variously referred to as *regulatory science* or *related scientific activities (RSA)*. The Bush report states:

> The preparation of accurate topographic and geologic maps, the collection of meteorological data, the determination of physical and chemical constants, the description of species of animals, plants and minerals, the establishment of standards for hormones, drugs, and X-ray therapy; these and similar types of scientific work are here grouped together under the term background research. Such background research provides essential data for advances in both pure and applied science. (Bush 1945, 82)

The objective of this type of activity is not primarily to advance fundamental understanding of nature, nor to solve a specific problem, but rather (to use the official definition of RSA) to 'complement and extend R&D by contributing to the generation, dissemination and application of scientific and technological knowledge' (Statistics Canada 2006a). As with applied research, the research objectives and methods to be employed are reasonably clear prior to the start of an investigation. This allows for comprehensive programs to be planned involving the coordination of relatively larger numbers of personnel. Regarding the appropriate organizational setting for this third type of scientific activity, the Bush report observes:

> There seems to be little disagreement with the view that these surveys and descriptions of basic facts and the determination of standards are proper fields for Government action and that centralization of certain aspects of this work in Federal laboratories carries many advantages. (Bush 1945, 82)

As indicated, the Bush report has been highly influential. Its categorization of scientific activity was later essentially codified by the OECD in the Frascati Manual, which, with modifications over the years, provides guidelines for how countries should report their national R&D data.[5] More importantly for the present purpose, the Bush report's arguments for particular organizational settings led to well-entrenched stereotypes about the appropriate locations for research activity. As Crow and Bozeman have summarized it:

> Universities are seen as the bastion of fundamental research, industry as the home of commercially related, applied and development research, and while the government lab stereotype is a bit more murky, government labs are often viewed as sites for supporting national research missions, especially in weapons, energy, space and agriculture. (1998, 14)

However, the authors go on to argue that resorting to stereotypes that assign particular categories of R&D activity to a sector of ownership is no longer adequate for answering the question of 'Where public science?' Linkages across sectors and among networks of laboratories are becoming increasingly evident and important (Gibbons et al. 1994). Some labs are 'hybrids' where public and private ownership and sources of funding are so complicated as to defy classification by simplistic sector-based categories (Crow and Bozeman 1998).[6]

Thus, while Bush's touchstone report helps us understand the historical rationales for locating public science, in particular institutional settings, its treatment is increasingly dated. Throughout the world, as the conduct of research has become more multidisciplinary, as research capacity has

---

5 The OECD, through the Network of Experts on S&T Indicators and others, is working to update its approach to reflect the evolving scientific enterprise.
6 This is the case in many Canadian labs, not only the so-called 'GOCO' laboratories that are government-owned but contractor-operated.

become more dispersed, and as knowledge production has become more distributed, the extent of partnerships and linkages within innovation systems and among public sector laboratories has increased. And in Canada there are a number of additional factors that are driving further integration of public science performers.

Contemporary Drivers for Co-Location

Achieving greater synergies within public science through the co-location of government labs with universities is not a new phenomenon. In Canada, the practice dates to before Confederation, when what would become the Meteorological Service of Canada located a facility on the University of Toronto campus. Today, there is an increasing trend toward co-location and there are a variety of factors contributing to the current interest. These factors operate at different levels of influence, from the political level of shifting views about the role of the state to the more mundane level of real property managers trying to solve immediate infrastructure concerns.

Alternative Service Delivery

A key driver behind co-location is the set of policy pressures related to alternative service delivery. As indicated in chapter 1, by the 1960s and 1970s there was increasing questioning of the appropriate model for public science. However, this debate seems to have been shaped by quite narrowly drawn dichotomies: public versus private, in-house versus ex-house, make or buy. Policy failed to appreciate the full spectrum of possibilities for organizing public science, ranging from fully public to fully private. By the 1980s and 1990s, administrative reforms associated with New Public Management continued to gain momentum and countries such as the U.K., Australia, and New Zealand altered their public science structures quite profoundly, including varying degrees of outright privatization of government laboratories.

Canada's public labs have similarly been under pressure both to become more market-oriented in their outputs and to adopt new institutional arrangements and management practices (Dufour and de la Mothe 2001; Doern and Kinder 2007). The latest political interest in alternative delivery models came in the government's science and technology strategy, *Mobilizing Science and Technology to Canada's Advantage*, released in May 2007. In it the government committed to considering 'transferring the management of some non-regulatory federal laboratories to universities in order to lever university and private sector strengths, create better learning opportunities for students and foster research excellence' (Government of Canada 2007, 70). While not a call for co-location per se, the commitment indicates a desire to explore alternative management arrangements and increase the linkage between government labs and universities.

Regional Innovation Systems/Technology Clusters

Another major driver towards co-location is the evolving understanding of the role played by public research institutions in fostering innovation and economic competitiveness, particularly at the local and regional levels. Building on the work of Marshall and other theorists over the last century, 'cluster theory' has emerged since the early 1990s as an increasingly popular policy tool (OECD 2001).

Developing internationally recognized technology clusters is one of the National Research Council's ongoing priorities, and NRC research institutes play a key role in these clusters (NRC 2006a). But the 'clusters' driver is not limited to the NRC. The CANMET Materials Technology Laboratory of Natural Resources Canada is being relocated to new state-of-the-art facilities at McMaster University. This move is motivated at least in part by a desire to build on synergies with local industry and universities 'to anchor a world-class innovation cluster supporting the needs of the energy, transportation, and manufacturing sectors' (Natural Resources Canada 2005). In addition, there is increasing interest on the part of provincial and municipal governments in the support of research parks and the clustering of R&D capacity.

More research is needed to understand the roles played in clusters by government laboratories, specifically. However, empirical work demonstrates that publicly funded research institutions, generally, play a role in almost all of the two dozen Canadian clusters studied by the Innovation Systems Research Network (Wolfe and Lucas 2005). These institutions are a critical component of the supporting infrastructure and organizational environment for cluster firms, not only providing R&D and testing services, but also contributing to the supply of highly skilled personnel in the cluster, providing specialized equipment and facilities, and facilitating networking and collaborative interactions. The clustering of government labs with universities helps to build a critical mass of research capacity 'in areas of local and regional strength to foster economic growth and improve quality of life' (Cassidy et al. 2005, 5).

The Old Rust and Grey

A final set of drivers for co-location emerges from the current state of the federal science and technology (S&T) enterprise. Given the R&D funding trends and for other reasons, federal labs face major challenges with respect to renewing their aging human resources and research infrastructures. Today, it can be said that the team colours of the federal S&T community are 'rust and grey,' as labs face a 'rust out' of their capital assets and a 'greying' of their workforce.

Through the lean years of the 1990s, capital investment in federal S&T infrastructure lagged as many labs tried to partially deflect the impacts of Program Review cuts (Doern and Kinder 2007). When the government did begin to reinvest in university research infrastructure through the Canada Foundation for Innovation federal labs were not able to access this funding.[7] More recently, the Treasury Board Secretariat led the Federal Laboratory Infrastructure Project (FLIP) to develop baseline data on the current state of the federal S&T asset base. According to one media report, FLIP was also about examining 'various models for governing federal S&T infrastructure, ranging from lab clustering with other federal labs, with universities and with industry' (Research Money 2007).

The other major issue related to the federal government's intramural capacity is the aging of its workforce. Just as with the university sector, the federal government has a large cohort of researchers nearing retirement and is in direct competition with universities for the next generation of researchers to refresh their ranks. Co-location is viewed as one way to increase visibility and attract students and post-doctoral researchers into government science.

A related driver comes from a long-standing concern about the geographic centralization of the federal R&D capacity. In 2004, the National Capital Region received 47 per cent of the total federal intramural R&D expenditures and was the location of 47 per cent of the federal scientific and professional personnel performing R&D activities (Statistics Canada 2006c). This pattern of concentration obviously varies by department, but politicians seeking to secure their region's piece of the federal pie are clearly interested in seeing more of this capacity decentralized throughout Canada. While this is not a driver for co-location per se, it does provide a pressure for the re-location of

---

7 The *Financial Administration Act* imposes constraints on the flows of federal funding.

government labs, and with this comes the potential that they will be moved to universities.

Case Study Labs

In this section I profile three examples of government laboratories co-located with universities. As indicated earlier, other examples could have been chosen. The three cases provide a mix of parent departments, host universities, research functions, and rationales for co-location.

Saskatoon Research Centre (SRC) at the University of Saskatchewan

The co-location of research centres with universities has been a long-established practice at Agriculture and Agri-Food Canada (AAFC).[8] Here, I profile the Saskatoon Research Centre at the University of Saskatchewan. It was selected as an early example of co-location, and as such makes clear that, despite the current buzz in Ottawa's policy circles, the practice of co-location in fact has a long history in Canada.

The Saskatoon Research Centre is located on the campus near Innovation Place, a research park specializing in agricultural biotechnology. The SRC is one of the department's nineteen research centres and it supports the departmental research mission to improve the long-term competitiveness of the Canadian agri-food sector through the development and transfer of innovative technologies. The specific mandate of the SRC and its three research farms is to 'bring a long-term commitment in crops research to the agri-food industry in Western Canada' (AAFC n.d.). With a staff of over 350, the SRC's key deliverables include improved germplasm of oilseed and forage crops; crop production and pest control practices; processes and products that expand the utilization of Prairie crops; and conservation, documentation, and distribution of diverse germplasm by Plant Gene Resources of Canada, Canada's main repository for seed.

The presence of the federal agriculture department in Saskatchewan dates to the earliest days following its establishment as a province in 1905, as research stations, experimental farms, and tree nurseries grew up in Indian Head, Rosthern, Scott, Swift Current, Melfort, Regina, and

---

[8] Examples include the Cereal Research Centre at the University of Manitoba, the Food Research Centre at the University of Guelph, and the Southern Crop Protection and Food Research Centre at the University of Western Ontario.

Sutherland (Dick 1996; Harding 1986). The roots of the Saskatoon Research Centre can be traced to 1917, when the first federal research officer was located at the university to 'work on the black fly and grasshopper problems facing the settlers' (Harding 1986, v). As the SRC's official history relates, many areas of Saskatchewan were plagued in this period by biting flies – mosquitos, horse flies, and black flies – with massive outbreaks causing a large number of livestock fatalities. With no professional entomologist in Saskatoon at the time, C.G. Hewitt, the Dominion Entomologist, recruited Alfred E. Cameron from British Columbia, where he had been fighting insect pests in fruit tree orchards.

According to Harding (1986, 3), 'the idea of having a resident entomologist working on a typically Saskatoon problem, i.e., black flies, also appealed to Dr. Walter Murray, President of the University.' Thus, in 1917, Cameron established the Dominion Entomological Laboratory on the campus of the young university. He was given a dual appointment and taught courses in entomology and parasitology in addition to his government responsibilities. Three years later, however, he resigned his federal post to take up full-time teaching at the university. This is an early example of one of the potential risks of co-location – the loss of key federal researchers to the university host.[9]

Another early challenge was the lack of stable laboratory space. The Lab spent its first five years in the basement of the Physics Building, followed by seven years in the College of Agriculture Building, and then seven years in the basement of the Field Husbandry Building. In 1936, it was moved into some offices rented in downtown Saskatoon. In 1938, the Lab returned to campus and for two decades was housed in the basement of the newly extended Field Husbandry Building.

Meanwhile, the Entomological Lab was joined on campus in 1919 by the Dominion Laboratory for Plant Pathology. The DLPP was established in the Horticulture Building and had one greenhouse. In 1931, the federal department established a third lab on campus, the Dominion Forage Crops Laboratory. The staff of this lab were federal employees but they had responsibility for all forage research and teaching on campus. In exchange, the university provided land, a greenhouse, and office

---

9 On the other hand, the lab's second director K.M. King not only stayed on for nearly a quarter-century (1922–46), he also found time to coach the University of Saskatchewan basketball team, once winning the Western Inter-Collegiate championship (Harding 1986).

and lab space for one dollar per year. In 1948, these three Agriculture labs were also joined on campus by the NRC's Prairie Regional Laboratory (now the Plant Biotechnology Institute). Their combined presence has been instrumental in fostering the major agbiotechnology cluster that exists in Saskatoon today.

Shortly following World War II, planning was begun to rationalize the three Agriculture labs into a single organization, the Saskatoon Scientific Research Laboratory, with its own building and supporting facilities. According to Harding (1986), the university was keen to keep the federal lab located on campus. Construction was delayed by the Korean War and other challenges but the building was finally occupied in 1957. The three Dominion labs became sections reporting to a single lab director. In 1959, the Lab became the Saskatoon Research Station and is now known as the Saskatoon Research Centre.

Interestingly, for most of its early history, the SRC was organized around traditional academic disciplines, i.e., entomology, plant pathology, agronomy, etc. Following a reorganization in the early 1980s, the lab is organized by crops – oilseeds, forages, cereals, etc. This is consistent with a general trend observed among government labs, shifting away from discipline-based organizational structures to more application-oriented structures (Doern and Kinder 2007).

In describing the benefits of the SRC's co-location with the university, the official history emphasizes the research and training synergies:

> The benefits of being located on campus have been twofold. Many graduate students, several of whom are now employed by the Research Branch [of AAFC], contributed to our research programs by doing their thesis research in our facilities under the supervision of our staff. Also, cooperative research with the University and with other research organizations on campus, such as the National Research Council of Canada, has been facilitated. (Harding 1986, v)

In addition, the early precedent of Cameron's cross-appointment continues to this day with federal researchers serving as adjunct professors.

National Wildlife Research Centre (NWRC) at Carleton University

The second case laboratory is the National Wildlife Research Centre (NWRC). Among federal science-based departments, Environment

Canada has been out front in experimenting with collaborative arrangements including co-location (Environment Canada 2004).[10] The NWRC is a particularly interesting case in that it previously existed as a standalone federal establishment that was then re-located to Carleton University. According to the NWRC, this location 'opens the door to new partnerships, giving government and university scientists new opportunities to collaborate on science that is critical to wildlife conservation' (Environment Canada n.d.).

For most of its history the NWRC was located in Hull, Quebec, on parkland originally bought by Senator Richard Scott. Built in 1863, Scott's home still stands on the property. The site was used by the Department of Agriculture's Animal Diseases Research Institute from 1918 until 1975, when the Institute was moved across the river to Nepean, Ontario. Environment Canada took over the site in 1976 and created the NWRC as the research headquarters of the Canadian Wildlife Service. In November 2002, the Centre moved to new research facilities on Carleton's campus.

With a staff of more than fifty people, the NWRC's mission is: 'To be the principal source of knowledge and expertise in the federal government on the impact of toxic substances on wildlife and the use of wildlife as indicators of environmental quality, and to conduct national surveys and research on migratory birds' (Environment Canada n.d.). Its activities centre on support for wildlife conservation and management activities; for regulation and compliance activities; and for decision making, understanding ecosystem functioning, and support for sustainable uses of the environment. It is important to note that the lab encompasses both R&D and related scientific activities. The NWRC exhibits very little in the way of commercial orientation and has very few links to the private sector. This is not surprising given the lab's mandate to deliver science as a public good and the relative lack of any commercial interests in wildlife research.

Given this lack of commercial focus, the NWRC co-location can not be viewed as part of a technology cluster, as is the case for the other labs profiled. On the other hand, the co-location can be seen as performing an analogous function by creating a central node of a research network.

---

10 Other examples from this department include the Meteorological Service of Canada's Canadian Centre for Climate Modelling and Analysis at the University of Victoria and the National Hydrology Research Centre at the University of Saskatchewan, and there are others.

'The close association between the NWRC and Carleton will act as the "hub" of a nationwide network of wildlife conservation science ...' (Turnbull 2004, 1).

A major driver for co-location was the previous state of the NWRC's physical assets, 'the oldest and arguably most out-of-date facilities in the Environment Canada science establishment' (Doern 2000, 300). Health Canada had identified a number of occupational health and safety problems that were threatening to close the Centre (Environment Canada 2004). In addition, the loss of core capacity in the 1990s coupled with the lack of adequate laboratory facilities were limiting productivity and research effectiveness. Studies indicated that the cost of renovations would likely exceed the cost of building new. Following an analysis of options, the department chose to pursue co-location. Although both the University of Ottawa and Carleton were approached, only Carleton put forward a proposal. The management team realized that 'by co-locating with a university that also had wildlife research strengths, it might be possible through enhanced collaboration to rebuild the critical research mass that NWRC had lost over the years' (Environment Canada 2004, 2). The NWRC faced an aging workforce and saw co-location as an avenue to attract more young people to the lab and into wildlife science more generally.

In the lab's new location it faces a number of key challenges and opportunities. Of particular interest is the impact that its location on campus is having on the lab's organizational culture and research effectiveness. The university is technically only its landlord (under a lease-to-own arrangement), but the co-location is changing the NWRC's day-to-day interactions. While the lab must continue to serve the policy and regulatory needs of Environment Canada, it now interacts with research faculty and students much more intensively. Collaborations have increased as more lab staff have become adjunct professors, sit on graduate thesis committees, and involve students in their own research (Grant 2004).

There is no question that the relocation solved many of the Centre's research infrastructure problems. The modern facility includes fifteen well-equipped laboratories, a greenhouse, and plant growth chambers for studies requiring controlled environmental conditions, and wildlife specimen storage facilities with three walk-in freezer rooms, four ultra-low-temperature chest freezers, and four liquid nitrogen storage banks.

A further example of positive collaboration came with the opening in 2004 of a $2 million Geomatics and Landscape Ecology Research

Lab. The Lab involves both Carleton faculty and NWRC researchers in work on remote sensing, species conservation, habitat mapping, biodiversity, forest structure, and population modelling (Carleton University 2004). This project also brought levered resources to the NWRC through the project's joint funding by the CFI, the Ontario Innovation Trust, and private donors, sources which are not directly accessible by the government laboratory but which are by the university (Doern and Kinder 2007).

On the other hand, there have been some challenges with the co-location. Due to Treasury Board rules regarding space allocation in government buildings, there is already a shortage of space for visiting researchers, post-doctoral fellows, and students (Grant 2004). There have also been issues related to building security and chain-of-evidence requirements which adversely impact the free access of researchers who are not federal employees (Environment Canada 2006). And there are indications that the lab, rather than escaping bureaucracy, now finds itself contending with two hierarchies simultaneously – the department's and the university's.

A more fundamental problem emerged from the initial overriding focus on the real property side of the relationship. The department is quite frank in its acknowledgement that the co-location did not adequately plan and fund research collaboration:

> Had the project been seen from the start as an integrated activity combining facilities and knowledge generation, then some of the important add-ons – research chairs, collaborative projects, student places, etc. – would have been fully funded as part of the original planning. By treating the research collaborations separately from the physical construction then some of the potential synergy was lost – or at least delayed to a future date. (Environment Canada 2004, 3–4)

However, based on interviews, the NWRC/Carleton co-location is viewed on the whole as a successful experiment and a model for future collaboration.

National Institute for Nanotechnology (NINT) at the University of Alberta

The final case profile is the National Institute for Nanotechnology (NINT) located on the University of Alberta campus in Edmonton. This institute is generating a great deal of policy interest as a potential model for federal laboratories and increasing government–university collaboration. More than mere co-location, the NINT represents a much higher degree of integration between government and university and is an exciting experiment with co-governance.

Nanotechnology research involves controlling matter at the molecular and atomic level to understand the characteristics and behaviour of materials, devices, and systems. It is viewed by many as 'the foundation of the next major revolution in technology, with societal and economic impacts expected by proponents to be larger than those of the computer revolution' (Rank et al. 2006, 8). At the very least, nanotechnology is expected to have major impacts across a wide range of applications in health, energy, biotechnology, high-performance materials, manufacturing, computing and information technologies, and consumer products (NRC 2001).

Seeking to become a world-class centre of expertise in this important emerging area, NINT is a unique joint initiative between the NRC, the University of Alberta, and the province of Alberta. Announced in August 2001, it was funded initially at $120 million, with the province and university together committing $60 million over five years and the NRC matching their investment with $60 million. The NRC has further pledged $12 million to support operating costs beginning in 2007–08. In addition, the federal Western Economic Diversification agency has invested $3.8 million to complete the construction of a business incubator facility (NRC 2005).

The Institute's mission is: 'To create knowledge and support innovation in select areas of nanoscale science and technology that will have long-term relevance and lasting value for Alberta and for Canada. NINT will employ a unique, collaborative, and interdisciplinary approach with exceptional research and support staff equipped with outstanding tools and facilities shared between NRC and the University of Alberta' (NRC n.d.). The new facility provides optimal nanoscale research conditions through clean rooms and 'quiet lab' space with ultra low vibration, minimal acoustic noise and electro-magnetic interference, and constant temperature and humidity (NRC 2006b).

NINT is intended to be a core element in the emerging nanotechnology cluster in Edmonton. In their evaluation, Rank et al. (2006) have identified a variety of factors that influenced the selection of Edmonton as the Institute's home. Recalling that the NRC seeks to establish clusters that build on local and regional S&T-based strengths, one may question why Edmonton was selected. According to the Canadian

Nanobusiness Alliance, Canadian nanotechnology clusters currently exist in five 'hubs': Montreal (with 40 per cent of the national nano-related organizations), Toronto (with 25 per cent), Ottawa (with 15 per cent) and Edmonton and Vancouver (each with only 10 per cent). Rank et al. (2006) suggest that Edmonton benefited from the fact that the University of Alberta had existing, perhaps unique, strengths and programs in nanotechnology-related areas such as protein science and information technology.

The evaluators also point out that nanotechnology has particular relevance to emerging sectors in Edmonton, including the biopharmaceutical sector, and the province is particularly interested in nanotechnology as it applies to the life sciences, the oil and gas industry, and information and communications technologies (Rank et al. 2006). Thus, NINT is well aligned with university and regional S&T and economic development priorities.

Finally, Rank et al. point to 'the commitment at senior levels among the three organizations to bring an NRC presence to Alberta, the only province without an NRC institute' (Rank et al. 2006, 10). This finding adds weight to the argument that the heavy geographic concentration of government laboratories in central Canada generally, and the National Capital Region in particular, is a consideration for decision-makers thinking about federal S&T policy.

The Institute's extensive governance structure reflects the high degree of integration among the three principals. Through the Institute's Director General, the NRC retains administrative reporting and legal accountability and is responsible for entering into all contractual, employee, and financing arrangements for NINT's operations. At the same time, this basic line authority is now paralleled by an Oversight Committee and Board of Trustees. The Oversight Committee, consisting of the NRC President, the President of the University, and the Deputy Minister of Alberta Innovation and Science, ensures that the strategic direction remains consistent with the priorities of the three principals. A Board of Trustees consisting of 12 external members and 4 ex officio members has only recently been established to review NINT's strategic plans and to monitor performance.

The NINT evaluation found that a significant but time-consuming achievement has been the negotiation and signing of several key agreements. These include the parent Memorandum of Understanding that provided the overall framework in which other agreements were to be negotiated, including agreements on governance, real property, intellectual property, secondments/cross-appointments, communications, and others. Significantly, the evaluation found that 'implementation has been somewhat slower than anticipated, due to the time needed to negotiate' these agreements (Rank et al. 2006, 48). Both sides of the partnership faced steep learning curves.

This early emphasis on building understanding of the needs and constraints of each partner seems to have paid off. Key outcomes in this initial phase include: a co-governance structure that reflects the interests of all partners, a unique physical infrastructure built with highly leveraged funds, and a blend of academic- and government-style research that addresses the strategic priorities of each partner. Of the 186 staff complement in October 2006, 55 were research staff, with 12 of these being cross-appointments from the university, and 106 were students and post-doctoral fellows. NINT has also attracted new talent, with 16 of the 23 principal investigators being new to Edmonton. Even in this early phase, NINT has had success promoting collaborations with large and small firms to the tune of $4.5 million in 2006–07 (Petersen 2007).

Co-Location Benefits and Challenges for Public Policy

In this section I begin to analyse the benefits and challenges associated with co-location from both the host university's and the federal government's perspectives. It must be reiterated that given the exploratory nature of this study the analysis is not exhaustive. For example, the perspectives of stakeholders beyond the federal government and host university are not considered. The issues raised here are intended to inform further investigation.

Benefits

For both sides of the co-location relationship, obvious key potential benefits derive from greater access to research knowledge, expertise, facilities, and equipment. For the university, joint appointments foster the cross-fertilization of ideas and can mean reduced teaching loads for university faculty. Co-location not only brings additional researchers to campus; it can also provide access to major, often unique, research facilities that the university would not typically secure on its own. While the CFI has been a critical source of funding for new research infrastructure, it has not provided much support for the operations and maintenance of that infrastructure (Schmidt 2007). A government lab, on the other hand, can not only bring unique equipment but also the well-trained technicians that operate and maintain it (Cassidy 2006).

There are also benefits to both sides related to research funding, although of a different nature for each side. As indicated at the outset, universities have been enjoying impressive gains in research funding over the past decade, while government labs have not. For universities, co-location is not viewed as a major source of new funding. On the other hand, a federal lab typically brings a more stable operating budget, in contrast to the more periodic nature of academic grant funding, and this facilitates more strategic planning and ongoing technical operations (Cassidy 2006). For the labs, the university hosts are typically not a direct source of funding, but cross-appointments can open new channels to the granting councils that would otherwise be off-limits to federal researchers.

Another area of mutual benefit is in the attraction and development of highly qualified personnel (Castonguay 2001). At NINT, the pooling of resources through the joint initiative has enabled the hiring of higher-quality researchers than would normally be possible (Rank et al. 2006, 32). By exposing students to more research and learning opportunities, co-location fosters increased integration of the university's research and educational missions. According to the NWRC study, one student has indicated that her decision to apply to graduate school at Carleton was influenced by 'the opportunity to study under Canadian government scientists who are leading experts on wildlife and environmental matters' (Turnbull 2003, 1).

Similarly, the government benefits when the presence of students advances the work of the lab, reduces its costs, and generates interest in government science as a potential career choice. At the SRC, 'one result of the university classroom involvement by staff members was that it made students aware of what the [lab] was doing.' This interaction helped recruit students for summer employment, which often led to research theses supervised by lab staff. These relationships 'increased considerably the lab's research productivity.' On the flip side, 'this preparation of future generations of research scientists, university lecturers and administrators was a major contribution of the [lab]' (Harding 1986, 27–8).

A final area of benefit to the host university that emerged from the study relates to reputation. As discussed elsewhere in this book, we are in an era of increasing national and even international competition among institutions of higher education. The major demographic shifts emerging as the baby boomers retire and the increasing mobility of the 'creative class' are creating fierce competition among universities for the best and the brightest, both among students and faculty researchers. Finding it more difficult to be all things to all people, universities are looking for ways to set themselves apart, to develop 'niches of excellence.' Having a federal laboratory on campus can bring added 'brand' attractiveness and reputational advantage.

And a final area of potential benefit to the government relates to priorities for public science. As the R&D funding trends continue to favour universities, federal departments and agencies will be increasingly challenged to deliver S&T in support of their public-good mandates. The focus is shifting from rebuilding internal capacity to ensuring access to the S&T capacity the government requires, wherever it resides. However, as discussed in chapter 5 in the analysis of the CFI, it is not at all clear what definition of 'capacity' is to apply here. The CFI's general definitions? The government lab's definitions? Are the government's policy and regulatory science capacity needs just an accidental add-on or are they precisely defined and negotiated as central parts of co-location decisions and arrangements?

This raises the related question of how the government might influence evolving academic research agendas to more directly support the nation's policy and regulatory needs, while maintaining academic independence and integrity. It is recognized that Bush's call for the 'free play of free intellects' still has merit and that top-down priority setting for academic science can be counter-productive. Co-location may foster a more bottom-up approach by helping to reduce the sectoral distinctions within public science and, over time, increasing the stake of university researchers in federal S&T priorities and issues.

Challenges

While the benefits of co-location are many, the case studies and literature also reveal some challenges associated with the practice. Decision-makers contemplating co-location arrangements should consider these challenges carefully.

Perhaps the biggest challenges are in the area of organizational culture and fit. As discussed earlier, the sectoral breakdown of research activity espoused in the Bush report is overly simplistic for the more complex S&T system that exists today. Nonetheless, it remains the case that there are important differences in the organizational cultures and processes found in universities and government labs:

> The federal government's research activities and those of universities fulfil quite different roles and there are certainly important differences in their approaches – e.g. the universities' much greater emphasis on peer review to ensure research quality, and on publication in peer-reviewed journals to disseminate research results. (AUCC 2007, 25)

The potential clash of cultures and differing norms around, for example, intellectual property, media communications, outside consulting, and career paths can negatively impact the partnership and reduce research productivity. It should be noted, however, that the AUCC goes on to state that there are also real possibilities for research collaboration beneficial to both sides. 'The extent to which federal research activities are already being conducted on campuses across the country is a clear sign that both parties see these partnerships as beneficial' (AUCC 2007, 25).

Another challenge relates to the provision and translation of science for policy needs. Ensuring effective communications across the science/policy interface can be difficult under any organizational arrangement. In a context in which enhancing accountability and value for money are core principles, a key question is whether the further organizational distance of a lab from its parent department that co-location represents will exacerbate this challenge.

A final set of challenges revolves around the identification and selection of a university host. As we have seen from the case profiles, the rationales for co-location and the decisions about with whom to co-locate differ considerably with each lab. With the SRC the decision was to create a new lab to address particular local problems in Saskatchewan and thus the choice of university host was straightforward. With NINT, a new national institute anchoring a nanotechnology cluster could have been located in many cities, and indeed strong arguments for other university hosts could be made. Finally, with the NWRC, an existing lab was re-located within the NCR through a process of invited bidding involving just two universities. In the absence of an overall policy relating to co-location, these choices were made in an ad hoc manner by the parent departments according to the logic of the particular situation.

However, if the government moves to a more intensive policy of partnering or transferring government laboratories to universities, questions arise about how candidate labs and host universities will be paired. Public policy questions that will need to be addressed include: Will the selection process be open for competition? Would NAFTA require that, for example, U.S. universities or university consortia be allowed to compete for the management of Canadian labs? Many questions remain.

Conclusions

Scholarship around, and policy interest in, public science and the government–university relationship has tended to focus on issues of governance and finance. For example, other authors in this book examine federal–provincial relations, the granting councils, and the regulation of research ethics. Less often examined are the issues around the *performance* of public science and institutional choice. This chapter has taken an exploratory look into this area by examining the co-location of federal labs with university campuses as a particular delivery mechanism for public science.

An initial observation is that, while co-location is receiving a considerable amount of 'buzz' these days, the practice is not new in Canada. The AUCC survey identifies more than 70 examples of federal/university research collaborations, some of which include co-locations (AUCC 2007). Current policy thinking can and should be informed by greater understanding of the partnering arrangements that already exist in Canadian public science.

Similarly, Canada is not alone in pursuing alternative delivery arrangements for public science and there are lessons to be learned from other jurisdictions. Other countries, including the U.S., the U.K., Australia, and Denmark, have adopted alternative arrangements within their public science systems. The Australian Government reviewed various practices critical to collaboration, including co-location, cluster formation, international and national networking, sharing of infrastructure, and co-investment in infrastructure and research. Interestingly, the review states that 'of these, the Committee considers that co-location is one of the most effective drivers of collaboration' (DEST 2004, ix). A systematic analysis of international practices with respect to co-location and related approaches would be timely.

Is co-location a model for the future? The answer is a qualified yes. Clearly it is no panacea and policymakers must guard against applying it as a 'one size fits all' solution. There are many legitimate reasons for some government science to continue to be performed in more isolated settings, whether for reasons of national security, access to particular environments, or to protect public confidence.

In addition, it must be recognized that there are many barriers to effective research collaboration (Government of Canada 2006). Co-location in and of itself is no guarantee that all of the potential synergistic benefits will be realized:

> Physical proximity does not by itself generate effective interaction, nor does it guarantee collaboration. This is because innovation performance also relies on the willingness and ability of organisations to interact, share and exchange knowledge. Co-location sets the scene for collaboration but requires much effort and nurturing of trust to forge the necessary links, build interactions and foster cooperation. (DEST 2004, 6)

Co-location is thus neither necessary nor sufficient for research collaboration, and achieving the full benefits of co-location requires concerted effort.

On the whole, however, it would seem that co-location is a model of public science delivery worthy of much greater consideration in Canada. The benefits flowing from the critical mass of expertise, talent, infrastructure, and funding can help counter Canada's problems of too little capacity spread over too much geography. Building on the impressive gains in university research support over the last decade, co-location can help ensure that Canada remains strong across the entire public science system.

REFERENCES

AAFC. (n.d.). Agriculture and Afri-Food Canada. From: http://sci.agr.ca/saskatoon/index_e.htm (cited 13 February 2007).

AUCC. (2007). *AUCC Submission to the Minister of Industry and the Minister of Finance on the Development of a Science and Technology Strategy for Canada.* Association of Universities and Colleges of Canada.

Bush, Vannevar. (1945). *Science: The Endless Frontier.* National Science Foundation, reprint 1990.

Carleton University. (2004). 'Carleton University Celebrates Opening of $2M Geomatics and Ecology Lab.' News release, 3 November.

Cassidy, Erin. (2006). Personal communications, 26 October and 28 November.

Cassidy, Erin, et al. (2005). 'Measuring the National Research Council's Technology Cluster Initiatives.' Paper presented at the CRIC Cluster

Conference: 'Beyond Clusters: Current Practice and Future Strategies,' Ballarat, 30 June–1 July.

Castonguay, Stephane. (2001). 'The Emergence of Research Specialties in Economic Entomology in Canadian Government Laboratories after World War II.' *HSPS: Historical Studies in the Physical and Biological Sciences* 32 (1): 19–40.

Crow, Michael, and Barry Bozeman. (1998). *Limited by Design: R&D Laboratories in the U.S. National Innovation System*. New York: Columbia University Press.

de la Mothe, John. (2000). 'Government Science and the Public Interest.' In *Risky Business: Canada's Changing Science-Based Policy and Regulatory Regime*, ed. G. Bruce Doern and Ted Reed, 31–48. Toronto: University of Toronto Press.

DEST. (2004). *Review of Closer Collaboration Between Universities and Major Publicly Funded Research Agencies*. Australian Government, Department of Education, Science and Training.

Dick, Lyle. (1996). 'The Greening of the West: Horticulture on the Canadian Prairies, 1870–1930.' *Manitoba History* (The Manitoba Historical Society) 31 (Spring): 22–33.

Doern, G. Bruce. (2000). 'Patient Science versus Science on Demand: The Stretching of Green Science at Environment Canada.' In *Risky Business: Canada's Changing Science-Based Policy and Regulatory Regime*, ed. G. Bruce Doern and Ted Reed, 286–306. Toronto: University of Toronto Press.

Doern, G. Bruce, and Jeffrey S. Kinder. (2007). *Strategic Science in the Public Interest: Canada's Government Laboratories and Science-Based Agencies*. Toronto: University of Toronto Press.

Dufour, Paul, and John de la Mothe. (2001). 'Change, Reform and Capacity: A Review of the Canadian Government R&D Experience.' In *Government Laboratories: Transition and Transformation*, ed. Cox et al., 97–113. Amsterdam: IOS Press.

Environment Canada. (n.d.). From: http://www.ec.gc.ca/scitech/default.asp?lang=En&n=A8279FD1-1 (cited 13 February 2007).

– (2004). 'A New Lease on (Wild)life – National Wildlife Research Centre.' *Smart Partners: Innovations in Environment Canada – University Research Relationships*. Environment Canada, Science Policy Branch.

– (2006). *Evaluation of the Co-location of Science Research Centres on University Campuses*. Prepared by Evaluation Division, Audit and Evaluation Branch, May.

Gibbons, Michael et al. (1994). *The New Production of Knowledge*. London: Sage Publications.

Government of Canada. (2006). *Overcoming Barriers to S&T Collaboration: Steps Towards Greater Integration*. Ottawa: Canadian Food Inspection Agency.
– (2007). *Mobilizing Science and Technology to Canada's Advantage*. Ottawa: Industry Canada.
Grant, Andrea. (2004). 'The National Wildlife Research Centre: Changing Approaches to Managing Government Science.' Unpublished research paper, School of Public Policy and Administration, Carleton University.
Harding, Howard, ed. (1986). *Saskatoon Research Station: 1917–1985*. Saskatoon: Department of Agriculture Historical Series 20.
Kinder, Jeff. (2003). 'The Doubling of Government Science and Canada's Innovation Strategy.' In *How Ottawa Spends 2003-2004: Regime Change and Policy Shift*, ed. G. Bruce Doern, 204–20. Don Mills, Ont.: Oxford University Press.
NRC. (n.d.). From: http://nint-innt.nrc-cnrc.gc.ca/home/index_e.html (cited 13 February 2007).
– (2001). 'NRC, Alberta Set in Motion a World-Class Center for Nanotechnology.' NRC News Release, 14 November.
– (2005). 'NINT Innovation Centre Focused on Commercialization.' NRC News Release, 12 October.
– (2006a). *Vision 2006*. National Research Council Canada.
– (2006b). 'Flagship Nanotechnology Institute's New Home Features Canada's Quietest Space.' NRC News Release, 22 June.
Natural Resources Canada. (2005). 'Materials Research: Government of Canada to Invest in State-of-the-Art Facilities.' NRCan News Release 2005/49, 6 July.
OECD. (2001). *Innovative Clusters: Drivers of National Innovation Systems*. Organization for Economic Co-operation and Development.
Petersen, Nils O. (2007). 'The NINT Partnership: Building a Cohesive Cluster Community.' Presentation deck, dated 24 September.
Rank, Dennis, Adele Acheson, and Mark Roseman. (2006). *Evaluation of the National Institute of [sic] Nanotechnology (NINT)*. Bearing Point.
Research Money. (2007). 'Flurry of activity around federal labs suggests government getting ready to move on making management changes.' *Research Money*, 5 February.
Schmidt, Sarah. (2007). 'World-class labs need big bucks to carry on.' *Vancouver Sun*, 13 February.
Statistics Canada. (2006a). *Federal Government Expenditures on Scientific Activities, 2006/2007*. Statistics Canada, Catalogue 88-001-XIE, vol. 30, no. 6 (September).

- (2006b). *Estimates of Canadian Research and Development Expenditures (GERD) Canada, 1995–2006, and by Province, 1995–2003*. Statistics Canada, Catalogue 88F0006XIE, no. 9 (September).
- (2006c). *Provincial Distribution of Federal Expenditures and Personnel on Science and Technology, 2000/2001 to 2004/2005*. Statistics Canada, Catalogue 88F0006XIE, no. 12 (September).

Turnbull, Chris. (2004). 'Environment Minister Opens National Wildlife Research Centre.' Carleton NOW, 27 January. Available from: http://www.now.carleton.ca/2003-04/121.htm (accessed 23 November 2008).

Wolfe, David A., and Matthew Lucas, eds. (2005), *Global Networks and Local Linkages: The Paradox of Cluster Development in an Open Economy*. Montreal: McGill-Queen's University Press.

10 Universities and the Regulation of Research Ethics

KARINE LEVASSEUR[1]

This chapter examines how research ethics are regulated in Canadian universities and explores the effects of these rules. It begins with a discussion of the origins of research ethics, which are the 'normative expectations' of behaviour and action, designed for researchers to follow (Johnson and Altheide 2002, 64). In a crucial sense, research ethics were first a system of self regulation arising from within the social system of science. In the last few decades, however, federal requirements for the regulation of research ethics have grown via rules, guidelines, and standards established or reinforced by the granting agencies (singly and collectively), federal science-based departments such as Health Canada as regulators, and by universities themselves. Accordingly, this chapter draws out a further key and changing feature of federal research and innovation policy, revealed in this case through regulatory policy means but functioning as a further layer of rules on top of the foundation of self-regulation and professional research norms.

Research ethics are intended to prevent and confront various types of unethical behaviour such as the falsification of research, conflict of interest, breaches to privacy and confidentiality, and the infliction of harm on research subjects. They may also provide guidance on other ethical issues such as the type of research that is allowed, relationships between researchers, and the protection of academic freedom.[2] To understand the

---

[1] Thanks are owed to Bruce Doern, Chris Stoney, Jerome Doutriaux, Allan Tupper, Emily Grafton, and the anonymous reviewers for their valuable comments and assistance. The author also thanks Peter Monette and Thérèse De Groote for comments made on an earlier draft of the chapter.

[2] I am indebted to Jerome Doutriaux and Allan Tupper for this idea.

regulation of these research ethics, the chapter first maps out the key regulators of research ethics frameworks, norms, and values in Canadian universities to provide insight as to how the ethics regime, and the federal role in it, has changed. Subsequently, the chapter examines the effects of the ethics rules on universities. It then examines the broader governance system of research ethics for all researchers, including public and privately funded research, because there may be important implications for university researchers. In order to understand these effects, however, we must first understand the origins of the research ethics regime and subsequent regulatory changes.

Origins of Research Ethics

Research ethics standards have long been self-regulated. Researchers regulate themselves in much the same way that other professions, such as medicine, promote professional responsibility and values. As professionals, university researchers have a responsibility to protect the public interest. As Dinsdale argues, an important aspect of being a professional is the

> contract with society on which professionalism is based. That contract requires public trust, which in turn depends upon the integrity of individual professionals and their organizations ... Professions become de facto trustees of the public interest if those principles are maintained in the face of pressures from the state or the marketplace. Professions must protect not only vulnerable persons, but also vulnerable social values. If professions do not meet their responsibility, public criticism ensues and, in the case of medicine, characterizes physicians as servants of the state or tools of business. (2005, 80)

Prior to World War II, decisions about research ethics were left to the discretion of researchers because a formal research ethics system did not exist (Sykes 2005). The ethics regime was initiated shortly after World War II as a result of the Nazi human experiments and the resulting War Crimes Tribunal (Interagency Advisory Panel on Research Ethics n.d.). In 1947, the first international document related to research ethics, the *Nuremberg Code*, established ten ethics standards for experiments involving human subjects (for example, voluntary consent). The *Nuremberg Code* was influential and 'brought the issue of human experimentation to the forefront of public debate and influenced a series of international documents created to ensure that all countries respect adequate standards of human dignity' (Interagency Advisory Panel on Research Ethics n.d.). The *Declaration of Helsinki*, a response to the *Nuremberg Code*, was created in 1964 and outlines the standards for medical researchers.

Despite these international documents being in place, cases of ethical slippage still occurred, such as the Tuskegee Syphilis Study (1932–72) conducted by the U.S. Public Health Service.[3] The study consisted of 600 men: 400 men who had contracted syphilis and 200 men who served as the control group. The 400 men with syphilis were low-income African Americans and were not informed that they had syphilis. Despite the discovery of penicillin in the 1950s, the research team deliberately withheld medication in order to document the progression of the disease (Berg 2007; Gold 2003). In response to this unethical treatment, the U.S. National Commission for the Protection of Human Subjects of Biomedical and Behavioural Research produced the *Belmont Report* in 1979, which outlined three standards for researchers: 1) respect for persons (individuals are autonomous and vulnerable individuals need protection); 2) beneficence (promotion of well-being); and 3) justice (fairness of distribution of harms and benefits of research within society). The ethics standards contained in these three documents are referenced in Canadian guidelines (Interagency Advisory Panel on Research Ethics n.d.).

Regulators

A review of the framework for research ethics in Canadian universities reveals complex structures encompassing multiple actors with multiple jurisdictions. Hirtle (2003, 139–40)[4] notes that the ethics framework faced by individual university researchers depends on several factors:

---

3 There have been other cases of ethical slippage, such as the Willowbrook Studies (1956–72) in the U.S. in which physicians infected mentally delayed children with hepatitis in order to understand the disease's progression. Other examples of research that raised ethical concerns in the 1960s include Stanley Milgram's study on obedience and Laud Humphrey's study on casual homosexual contacts in public washroom facilities. More recently, ethical concerns have been raised in this country related to the possible pressure exerted by industry sponsors on researchers in light of the incident involving Dr Nancy Olivieri and Apotex, a pharmaceutical company (for details see Gibson et al. 2002; Schuchman 2005; Thompson et al. 2001).
4 To accord with the theme of this chapter, some of these factors outlined by Hirtle have been slightly altered.

- type of research (scientific and bio-medical research has more layers of review than other types of research);
- residency of the researcher (provinces and territories have different legislation such as privacy legislation);
- sources of funding (funders may have their own ethics standards); and
- category of research participant (human or animal subjects).

Given this complexity, mapping the *key* regulators of research ethics in Canadian universities becomes an important task for this chapter in order to understand changes to the regulation of research ethics. The chapter also reports on the effects of the changes to these rules on Canadian universities and their researchers.

While significant differences exist between the approaches to research ethics for human and animal research subjects, one shared feature is the use of Tri-Council funding to enforce research ethics principles (Schuppli and McDonald 2005, 97). The federal granting agencies set the terms and conditions for funding vis-à-vis the signed Memorandum of Understanding (MOU) that specifies the expectations of how research should occur. The federal granting agencies require universities to sign the MOU. The MOU contains an annex on ethics so that all research being done at Canadian universities must follow the *Tri-Council Policy Statement on Ethical Conduct for Research Involving Humans* (TCPS) and the Canadian Council on Animal Care (CCAC) standards for research involving animals. Compliance with research ethics standards in Canadian universities, therefore, is achieved through a 'contractual model' (Hirtle 2003, 141; Lemmens 2005; Schuppli and McDonald 2005, 97). Given the differences between research ethics for human subjects versus experimental animals, it is important to outline these different approaches.

Ethics for Research Involving Humans

Federal Granting Agencies

Prior to the adoption of the TCPS in 1998, the individual federal granting agencies had their own research ethics policies. Guidelines for bio-medical research ethics were first developed in the 1970s by the Medical Research Council (MRC)[5] and later revised in 1987, with the addition of

---

5 The Medical Research Council is now the Canadian Institutes of Health Research.

new guidelines in 1990 on Somatic Cell Gene Therapy (Interagency Advisory Panel on Research Ethics, n.d.). Comparatively, the Social Sciences and Humanities Research Council (SSHRC) borrowed the Canada Council Guidelines for research ethics, while the Natural Sciences and Engineering Research Council (NSERC) used SSHRC and MRC research ethics guidelines since this Council did not develop its own guidelines (Interagency Advisory Panel on Research Ethics n.d.).

The publication of the TCPS marks the unification of one set of ethics for the three federal granting agencies: Canadian Institutes of Health Research (CIHR), SSHRC, and NSERC. Three updates to the TCPS have been completed to date. The first update in May 2000 clarified that institutions may create an appeal board to review REB decisions 'provided that the board's membership and procedures meet the requirements of [the TCPS].' In addition, this update added a provision to help smaller institutions establish an appeals process and corrected errors in the French translation of the document. The second update in September 2002 clarified that ethics review is required in order to gain access to 'private papers' and added an explanatory text on maintaining privacy and confidentiality when working with databases.[6] The third update in October 2005 made 'editorial and technical corrections' (CIHR, NSERC, and SSHRC 1998/2005).

While the federal granting agencies provide a uniform standard for research ethics, the level of interest in research ethics is somewhat higher for the CIHR and SSHRC. Halliwell (n.d., 2) indicates that approximately 5 to 7 per cent of NSERC's constituency are affected by research ethics, so this granting council has 'modest interests' in the issue. For SSHRC, ethics is an important issue given that nearly 50 per cent of researchers in the social sciences and humanities require ethics approval. SSHRC must also deal with a diversified research community engaged in research with various levels of risk. To provide policy advice concerning research ethics from the social sciences and humanities perspective, SSHRC established the Standing Committee on Ethics and Integrity as part of its governance structure. The CIHR, given the high level of risk to research subjects, has a strong commitment to the development of research ethics and has established an active Standing Committee on Ethics to identify emerging ethical issues within health research. Additionally, the CIHR has an ethics office comprising approximately seven to nine staff members (Halliwell n.d.).

---

6 There was also a minor typographical correction in the second update.

The TCPS guidelines embody several guiding principles, including respect for human dignity in all of its means and ends. Research must achieve an acceptable end 'in terms of the benefits for subjects, for associated groups, and for the advancement of knowledge,' but must also employ ethically appropriate means to achieve its end results (CIHR, NSERC, and SSHRC 1998/2005, i.4). Other ethics principles embedded in the TCPS include respect for free and informed consent, respect for vulnerable persons, respect for justice and inclusiveness, and respect for privacy and confidentiality. Other guiding principles emphasize the need for research to minimize harms and maximize benefits. Additionally, research must carefully balance the distribution of harms and benefits.

The definition of 'research' determines when the TCPS policy must be invoked. Article 1.1(a) states, 'All research that involves living human subjects requires [ethics] review and approval by an REB in accordance with this Policy Statement, before the research is started' (CIHR, NSERC, and SSHRC 1998/2005). More specifically, research that uses 'human remains, cadavers, tissues, biological fluids, embryos or foetuses' requires review and approval by an REB prior to commencing the research project (Article 1.1[b]). Also, 'research about a living individual involved in the public arena, or about an artist, based exclusively on publicly available information, documents, records, works, performances, archival materials or third-party interviews' is exempt from ethics review and approval unless interviews or private papers are used (Article 1.1[c]). Quality assurance systems, performance reviews, or testing is not subject to ethics review (Article 1.1[d]).

The TCPS policy is intended to be an evolving document that responds to new demands. The federal granting agencies continue to be the stewards of the TCPS, but also established a multidisciplinary mechanism in 2001 known as the Interagency Advisory Panel on Research Ethics (PRE) to advise on ways in which the TCPS can evolve and meet the needs of researchers. The PRE also nurtures dialogue related to the development of a governance system for the ethical conduct of research involving humans. To advise the granting agencies, the PRE has working committees and organizes consultations to explore key issues related to research ethics. In 2006, the PRE held consultations on the interpretation of proportionate review and privacy/confidentiality. Consultations in 2007 explored ethical issues related to sharing information in clinical trials, Aboriginal research, issues faced by qualitative researchers who must abide by the TCPS, and refinements to continuing research ethics review.

For research involving humans, the TCPS guidelines are implemented by local, institutional Research Ethics Boards (REBs). These REBs, Hirtle notes, 'are considered the core structure of ethics review systems' for research involving human subjects (2003, 140). Article 1.3 of the TCPS specifies that REBs must consist of at least five members. Of these members, two must have expertise in methods or areas of research covered by the REB; one must be from the community with no affiliation to the institution; one must possess ethics expertise; and, for biomedical research, one must have legal knowledge.

REBs must approve all research involving human subjects as per the signed MOU since the TCPS standards apply to *all* research involving humans. REBs, therefore, are designed to ensure compliance with the TCPS. Researchers at Canadian universities are required to submit proposals for all research projects involving human subjects for review and approval. Through regularly scheduled face-to-face meetings (Article 1.7) backed by proper record-keeping procedures of meeting minutes (Article 1.8), REBs can approve, reject, propose modifications to, or terminate a research project that has already been initiated with ethics approval (Article 1.2). The TCPS outlines a three-tiered decision-making process that includes an initial review of the proposed research (Article 1.10).

This is followed by a reconsideration in which researchers have the right to request a reconsideration of decisions pertaining to the proposed research (Article 1.11). The third tier consists of an appeals process in which the TCPS allows for the establishment of an appeal board for situations in which members of the REB disagree, provided that the appeal board meets the specifications of the TCPS policy (Article 1.11). The university cannot override negative decisions of the REB so as to ensure the independence and authority of the REB, but the university can override positive REB decisions (Article 1.2).

The REB is very powerful and can do one of four things: it can approve a research proposal outright; approve a research proposal with modifications; reject a research proposal; or, terminate a research project that is already in progress. In instances where non-compliance is determined, the REB can withdraw its approval, which would effectively stop the research. REBs do not discipline researchers in instances where misconduct is believed to have occurred. Instead, the REB refers the matter to the institution, which must follow its integrity policy and investigate instances of misconduct and report its findings to the federal granting agencies. In order to receive funding from the federal granting agencies, all universities must have an integrity policy in place to: promote research and scholarship integrity; prevent misconduct; and, address misconduct (see 'Schedule 4: Integrity in Research and Scholarship of the MOU'). Additionally, some universities may have individual policies that regulate the topic of research that can be undertaken by staff, faculty, and students for ethical or political reasons (for example, universities may ban research related to weapons of mass destruction).[7]

Health Canada

Another change in the relationship between the federal government and Canadian universities with regards to research ethics relates to Health Canada and its role as the regulator of clinical research such as new drug trials and new medical devices. The Therapeutic Products Directorate (TPD) of the Health Products and Food Branch, housed in Health Canada, is the drug licensing authority that reviews and regulates new drugs and medical devices for safety. Health Canada does not regulate university research ethics per se. Rather, this federal government department regulates research that may be conducted at universities and university-affiliated hospitals with academics as the principle investigators.

The Food and Drugs Act regulates the sale of therapeutic products in Canada. The TPD endorses the Good Clinical Practice Guidelines published by the International Conference on Harmonization (ICH-GCP), which is an international standard-setting body for research into new drugs to ensure the well-being, safety, and rights of human subjects participating in clinical trials. These guidelines also attempt to streamline the drug approval process for regulatory authorities of Europe, Japan, and the United States (Lemmens 2005). While Canada has observer status and endorses the ICH-GCP, Canada is also an active participant in developing guidelines (Government of Canada, Health Canada 2002a).

In 2000, regulatory amendments to the *Food and Drug Regulations* were announced that would allow for inspections of clinical trials to begin effective 1 September 2001. To proceed with a clinical trial in light of these new amendments, drug sponsors, including universities and university-affiliated hospitals, must apply to and receive authorization from Health Canada. Upon authorization from Health Canada,

---

7 I am indebted to Jerome Doutriaux and for this idea.

sponsors must also identify a REB and obtain its approval for the conduct of a clinical trial and the use of informed consent forms (Government of Canada, Health Canada 2003). Inspections of clinical trials are conducted by the *Health Products and Food Branch Inspectorate* to ensure compliance with the ICH-GCP, Division 5 of the *Food and Drug Regulations*, and other relevant guidelines. These on-site intensive inspections are mostly announced inspections, although Health Canada has the authority to make unannounced visits as well at its discretion. If the inspection reveals non-compliance, the sponsor of the clinical trial has the opportunity to correct any weaknesses, although the *Health Products and Food Branch Inspectorate* may pursue other necessary action as per the *Compliance and Enforcement Policy* (Government of Canada, Health Canada 2002b, 7).

Research Ethics Involving Animals

In 1968, the Canadian Council on Animal Care (CCAC) was formed, and today it consists of over 2,000 volunteers mandated to oversee the development and implementation of standards related to animals in research (Gauthier 2004). To achieve its mandate, the CCAC operates three complimentary programs: guideline development for the care and use of animals; institutional assessments; and educational training/communication programs to support local Animal Care Committees (ACC).

The CCAC sets and implements the standards for the care and use of animals. The standards for research ethics involving animals cover such things as animal care, housing for animals, availability of water, food, exercise, social needs, and euthanasia. A shared feature of the ethics frameworks for research involving humans and animals involves the enforcement of standards through a contractual funding model (Shuppli and McDonald 2005). As noted earlier, to be eligible for funding, universities and other grant-receiving institutions must sign a MOU with the federal granting agencies that requires compliance with CCAC standards and the TCPS.

Another shared feature is the use of institutional review mechanisms (Shuppli and McDonald 2005). The CCAC established local ACCs to ensure compliance with the standards in each institution. ACCs, therefore, are the equivalent of REBs. Members of an ACC are appointed, and the composition of each ACC reflects the needs of the institution. Generally, ACCs are comprised of scientists and/or teachers, veterinarians, technical staff, students, members from faculties not using experimental animals, and members from the community (CCAC 2006). ACCs must review and approve research protocols involving animals before researchers can begin their project.

To ensure accountability and quality control, institutional assessments based on a peer review model are conducted at least once every three years (Schuppli and McDonald 2005). While these external assessments are conducted with the local ACCs and institutions, they also 'include the assessment of the functioning of the CCAC' (Gauthier 2004, 58). The assessment panel is comprised of scientific members, together with at least one veterinarian and a community representative. An assessment begins with a pre-assessment meeting of the panel members to 'review the goals of the assessment, the institution's pre-assessment documentation, the results of the previous assessment, and any other issues related to the visit' (CCAC 1999). This is followed by a meeting to review the implementation of previous CCAC recommendations, ACC documentation and protocol, and ACC site visits of animal facilities (CCAC 1999). A site visit then occurs to assess animal holding and care facilities, and any other areas in which animals are used for research purposes. Meetings and site visits occur with members of the ACC, senior animal care personnel, and senior institutional representatives. Upon completion, the panel prepares a formal assessment report that is sent to the institution within ten weeks. If no recommendations are contained in the assessment report, the Assessment Committee classifies the institution as operating in compliance with CCAC standards. The institution is then awarded a Certificate of Good Animal Practice®. Institutions that require changes to their protocols, as per the assessment report, are given opportunities to implement such changes. For those institutions that are still not determined to be in compliance, the CCAC must report them to the CIHR and NSERC as per the MOU. Research funding can be suspended from the federal granting agencies when such a breach is determined.

Emerging Regulators

Two emerging regulators of research ethics complement the self-regulatory roots of the system and federal rulemaking and guidelines via the granting agencies and departments such as Health Canada. These emerging regulators include international regulators and refereed journals. Foreign governments demand accountability vis-à-vis the establishment of appropriate safeguards for research subjects involved in research that is funded by a foreign government, but is conducted in Canada (Halliwell 2005). If research is conducted with funding from a foreign government agency, researchers must also comply with the ethical requirements of that country. For example, Canadian researchers who receive American funding through the National Institutes of Health must also ensure that their research protocols comply with American ethical and legal regulations for research involving humans (Schuppli and McDonald 2005, 100).

Refereed journals also have the capacity to reinforce research ethics. The International Committee of Medical Journal Editors (ICMJE), consisting of approximately a dozen bio-medical journals, is a prime example of how refereed journals can reinforce research ethics. In 2001, the ICMJE became concerned about the potential for the suppression of data with industry-sponsored research and opted to implement policies governing the publication of industry-sponsored research (Baird 2003). These new policies and rules require authors to verify that they had access to all of the research findings and to assume responsibility for the research integrity of the study and the results, and editors have the right to review research protocols and funding contracts when considering a paper for publication (CMAJ 2001). Likewise, authors must register their clinical trial with a public registry to be considered for publication (ICMJE 2007).

Effects of the Rules

As indicated, the research ethics regime began with researchers themselves in the form of self-regulation. In Canada, it was in the 1960s and 1970s that the federal granting agencies began regulating ethics for research involving animals and humans respectively. The federal government, which funds the federal granting agencies, which in turn fund research at Canadian universities, does not want human and animal research subjects to be harmed by research that is funded with public money. This need to be accountable for the use of public funds explains how the federal granting agencies collectively came into their role of setting ethics standards for all research in Canadian universities undertaken by faculty, staff, and students. Since then, the adoption of a unified set of research ethics in 1998 by the three granting agencies for research involving humans, coupled with the new clinical inspections undertaken by Health Canada in 2001 and the continued work of the

CCAC, signal significant federal regulatory change – with several effects for Canadian universities.

First, there are the effects of the 'one-size-fits-all' approach to the TCPS policy (Grayson and Myles 2004; van den Hoonaard et al. 2004). Although efficiency may be gained through the adoption of a single policy and set of requirements on research ethics (Kellner 2002), the regulatory regime may not be appropriate for all types of research (Hirtle 2003, 141; Lougheed 2003, 12). The TCPS, inspired by the biomedical model, may be applied in such a way that the TCPS 'does not "speak" to [the experiences of all researchers], leaving REBs that may lack appropriate breadth of expertise free to impose default assumptions that threaten free inquiry for no ethical gain' (van den Hoonaard et al. 2004, 10).

The TCPS allows for various forms of informed consent, but the application of the TCPS may at times largely focus on the use of signed informed consent forms.[8] Consent forms are contentious for some social science researchers because they may jeopardize anonymity, prevent some subjects from participating in the research, and undermine a friendly relationship between the researcher and subject (van den Hoonaard 2002, 10). Signed consent forms are particularly challenging for social science researchers employing a 'participant observation' methodology, in which the researcher observes and records the daily lives of members in a particular group. This mode of research requires the researcher to minimize his or her impact on the group, but the required use of signed consent forms may undermine this type of research 'because human "subjects" would regard the requirement of such a form as coercive' (van den Hoonaard 2002, 10).

The 'one-size-fits-all' approach of the TCPS, coupled with the fact that REBs are liable for negligence (Gold 2003), may lead some REBs to adopt a more legalistic, bureaucratic, and conservative stance to reviewing research proposals (Grayson and Myles 2004; McDonald 2005). Grayson and Myles share such an experience with their SSHRC-funded collaborative research project designed to identify the experiences of students in their first few years of study. Researchers at York University, the University of British Columbia, McGill University, and Dalhousie University worked collaboratively to survey students at each university and designed a standard letter of introduction. Since collaborative research must meet the standards of each university, the REBs at each of

---

8 Personal correspondence.

the four universities reviewed the uniform letter of introduction. The letter of introduction was approved outright by York's REB, and approved by Dalhousie's REB with only small changes required. McGill's REB also approved the letter, but required a consent form written in legalese. The REB at UBC rejected the letter, citing, also in legalese, the need for more details, which were subsequently supplied. The appropriate versions of the letter of introduction (and, at McGill, the consent form) were sent randomly to students at all four universities. Response rates were lowest at McGill University, followed by UBC. The impact, Grayson and Myles argue, is significant, in that

> inter-university research is constrained by the requirements of the most ethically conservative university. The net effect of these and other constraints is that in the humanities and social sciences Canadian scholars are frequently compelled to make what are from their point of view unwarranted changes in research topics and methodologies in order to receive REB approval. (2004, 294)

Nowhere is this concern more clearly outlined than in the report submitted to the PRE from the Social Sciences and Humanities Research Ethics Special Working Committee. Engaging in national consultations, this report, titled *Giving Voice to the Spectrum*, outlines the 'deleterious effects of the TCPS':

- some REBs, because of the biomedical model underpinning the TCPS, are not familiar with other research methods employed by social science and humanities researcher leaving them ill-equipped to review proposals using non-biomedical methodologies;
- some students have paid extra tuition because of prolonged ethics review and approval;
- some social science and humanities researchers have been told by REBs to avoid using certain methodologies that are problematic for ethics compliance (van den Hoonaard et al. 2004, 10).

These effects infringe upon academic freedom for some researchers. As van den Hoonaard et al. argue,

> The submissions suggest that the ability of social sciences and humanities researchers to engage in and fulfil their traditional mandate to gather information about and critically analyze all aspects of society is being

threatened by a narrowing of permissible topics and approaches that has nothing to do with 'ethics' and everything to do with non-ethics criteria such as liability management and other forms of 'ethics drift.' This has infringed on academic freedom. (2004, 11–12)

Miller's research into the development of SSHRC also acknowledges this concern. He notes that in the early 1990s there was strong pressure for the federal granting agencies to work collaboratively. The source of this pressure was the possibility that the federal government would move towards legislation to regulate research involving humans.[9] In response to this pressure, SSHRC opted to participate in discussions to create the current TCPS. The development of this uniform policy and regulatory statement on research ethics based on the biomedical model, however, has been problematic for SSHRC. Miller notes that the adoption of the TCPS has 'tied the hands of at least some researchers unnecessarily, imposed administrative burdens on universities, and still left an important, emerging area of research – research dealing with Aboriginal peoples – inadequately covered' (Miller 2006, 23).

This discussion raises the question as to whether researchers, research subjects, and academic freedom are better served by a 'one-size-fits-all' model of research ethics or by several models. The PRE conducted a consultation from February to April 2007, with a discussion paper as a follow-up to the *Giving Voice to the Spectrum* report to articulate policy recommendations. A linked prior question on these kinds of issues is whether there should be a federally mediated model at all. The answer is likely to be yes, largely because, as long as the federal government is a key research funder, it will want some assurance that research ethics constitute a central part of the contract between researchers and federal taxpayers.

Another effect of the rules relates to how these rules make Canadian universities 'internal regulators' of research ethics for faculty, staff, and students.[10] Besides regulating research, all university researchers, administrators, and offices of research services promote research integrity throughout the university.[11] 'Research ethics' is differentiated from 'research integrity,' with the latter being a much broader umbrella within

---

9 Personal correspondence.
10 This is not to suggest that universities would not have developed their own ethics guidelines in the absence of externally imposed standards.
11 The ideas contained in this paragraph are attributed to personal correspondence.

a university. Research integrity concerns the conduct of research overall in the university and the establishment of an organizational culture that is supportive of research ethics.

Concerns have been raised about the ability of REBs to undertake the role of an internal regulator on behalf of the federal granting agencies. REBs have been criticized for having membership and procedural irregularities and for being too bureaucratic (Baker n.d.; McDonald n.d.). Participation on a university REB, however, is a very challenging and demanding task. As McDonald explains,

> Generally, REB membership is regarded as an onerous, unrecognized assignment with no prestige. Most REBs run on a shoestring budget with slight institutional support at best. There is little in the way of ethics education for REB members and established or new researchers. (McDonald n.d.)

REBs are also hampered by their inability to communicate with other REBs. As Ferris notes,

> Whereas some REBs may communicate informally with one another about a protocol, such communication is sporadic at best. Such sharing of information may be hazardous, given the absence of clarity about industrial or intellectual property rights and legal direction about how much can be revealed and to whom ... We need ... mechanisms that will increase communication and coordination. (2002, 10)

REBs also have to contend with new funding sources of university research. Research is becoming increasingly complicated because of the recent introduction of industry sponsorship. The current research ethics regime for Canadian universities was established when public funding to universities was more dominant (Baird 2003). The result, she notes, is the increased pressure for university researchers to secure industry sponsorship. Conflicts may arise when research results are negative or potentially harmful to the sponsoring company. In recent years, there have been calls for REBs to review all research contracts for industry-sponsored research to prevent the suppression of data and protect the researchers (Gold et al. 2003, 13).

Monitoring research has also been identified as a concern regarding regulatory compliance. The TCPS (Article 1.13) specifies that REBs must review ongoing research, in which researchers must propose an appropriate process for ongoing review by the REB. At the very least, this must consist of the submission of an annual report. While REBs are responsible for monitoring ongoing research, McDonald notes that the 'lack of monitoring and auditing processes for ongoing and completed research' may prevent REBs from knowing what happens with research projects once they are approved (2005, 6). Weijer contends that 'few research ethics boards fulfil this responsibility' (2001, 1305). The National Council on Ethics in Human Research (NCEHR) notes that non-compliance is known to occur (2006). While site visits are provided at no cost to institutions (for example, universities, hospitals, community-based organizations, private organizations) by NCEHR, these assessments are voluntary. NCEHR's site assessment reports are confidential, but its *Task Force for the Development of an Accreditation System for Human Research Protection Programs* (2006, 10) reports that these site visits 'have revealed problematic practices in the oversight of research with human participants.'

There are also concerns that the federal granting agencies may be in a conflict of interest related to research involving human subjects. The granting agencies have dual roles as both promoters of *research* through their granting functions and protectors of *human subjects* through their regulatory roles that may prevent them from being viewed as independent, given that their primary role is to build knowledge and understanding (Schuppli and McDonald 2005). As Schuppli and McDonald (2005, 101) argue, 'the Councils cannot be seen as independent protectors. At best, they are "interested" or "invested" protectors who profess to have balanced the objectives of research promotion with human subject participation.' This raises the question as to whether a separate and independent organization operating at arm's length from the granting agencies is required to develop ethics standards. Comparatively, the potential for a conflict of interest for the federal granting agencies related to animal research subjects is limited. The conflict of interest is limited because the CCAC enjoys greater independence from the federal granting agencies and the pressure to promote research.

Schuppli and McDonald outline several organizational features that provide greater independence for the CCAC. First, the governance of the CCAC is diverse and includes 24 member organizations. Of these 24 member organizations, the federal granting agencies only have one vote each, meaning that the behaviour of the CCAC is not solely determined by the agencies. Second, the participation of animal advocates in the CCAC helps shield the ACCs and CCAC from the pressure to promote research. Third, standards for the care and use of experimental animals are developed by CCAC stakeholders through consultations 'which have not been subject to review or approval by the research councils' (Schuppli and McDonald 2005, 101).

Given the multiplicity of ethics regulators for research involving humans, the concern has been raised that the ethics regulatory regime is too complex. This may cause confusion as to 'which rules apply to which types of research [making it] unclear how to handle, for example, trans-provincial research or research with funds from the United States' (Hirtle 2003, 14). There also exists the possibility for conflicting ethics standards in light of multiple regulators. Bevan (2002) notes that in one application for ethics approval of a new drug trial, the REB, using the TCPS guidelines, was in direct conflict with the TPD, which endorses ICH-GCP guidelines. Increased compliance costs and delays may result from the presence of multiple regulators employing potentially conflicting standards (Schuppli and McDonald 2005). Besides increased compliance costs, the protection of human subjects may be jeopardized by these complex structures without the presence of a single, national oversight system.

While no single oversight system exists for the ethical conduct of *all* research involving humans, there have been calls for such a system to be established (Bevan 2002; House of Commons Standing Committee on Health 2004). Research involving human subjects at universities and other research institutions funded by the federal granting agencies is governed by the TCPS. However, a significant portion of research involving human subjects conducted at federal, private, and non-profit labs does not fall under the TCPS, although these labs may voluntarily opt to comply with the TCPS (NCEHR 2006; Gold et al. 2003, 3). This may result in uneven layers of protection for human subjects, since universities meet the standards of the TCPS, but 'when other research hosts and funders are involved such as government or community-based or purely private research, few, if any, research ethics structures exist' (Hirtle 2003, 140).

There has been increasing pressure from the U.S. to develop an accreditation system. The report titled 'The Globalization of Clinical Trials: A Growing Challenge in Protecting Human Subjects' indicates that more and more foreign clinical trials are being conducted in which the data is used by the Food and Drug Administration (FDA) to determine the safety and efficacy of new drugs to be marketed in the U.S. (Rehnquist, 2001). The FDA, therefore, approves drugs in which the data was collected in a foreign country, but it does not know how the research subjects were ethically treated. The FDA wants to ensure that the protection for research subjects in foreign clinical trials is equivalent to the protection for human subjects in American clinical trials. The recommendations stemming from this report make it very clear that the American government wants to see action on this front. A key recommendation of the report supports the development of an international accreditation system.

In response to the increased pressures for greater protection of human subjects, the Sponsor's Table for Human Research Participant Protection in Canada was organized.[12] The Sponsor's Table mandate centres upon developing an accreditation or alternative system for the protection of human research subjects. Accreditation has higher standards than regulations and is defined as 'a non-governmental, self-assessment and external peer assessment process used by professional organizations to accurately assess levels of performance in relation to established standards and to implement ways to continuously improve' (Dinsdale 2006, 4). The Sponsor's Table, through its Experts Committee, is still in the early stages of its work, so the type of system and its specific form remains to be seen; however, it is likely that more changes are forthcoming for all research institutions, including Canadian universities (Halliwell 2005).

Comparatively, the framework for the ethical care and use of animals in research is more precise because of its accreditation system through the use of the Good Animal Practice Certificates®, well-developed accountability and transparency structures, and limited conflict of interest (Schuppli and McDonald 2005, 97). Indeed, Canada's framework for the ethical treatment of experimental animals has received international praise. In 2003, the U.S. National Academy of Science remarked that the CCAC is the best framework for animal care standards in the world, and CCAC guidelines have been distributed in English, French, and Spanish for use worldwide (CCAC 2003–04; Schuppli and McDonald

---

12 The Sponsor's Table is comprised of: Alberta Ministry of Health and Wellness, Association of Canadian Academic Healthcare Organizations, Association of Faculties of Medicine of Canada, Association of Universities and Colleges of Canada, Canada's Research-Based Pharmaceutical Companies, Canadian Federation for the Humanities and Social Sciences, CIHR, Fonds de la recherche en santé du Québec, Health Canada, Health Charities Coalition of Canada, Michael Smith Foundation for Health Research, Research Canada, NSERC, SSHRC, and the Royal College of Physicians and Surgeons of Canada.

2005). Given the strengths of the CCAC model, there may be lessons learned that can be applied to the framework for research ethics involving human subjects.

Conclusions

This chapter has examined how research ethics are regulated in Canadian universities and has explored the effects of these rules. It has shown that research ethics were first a system of self-regulation arising from within the social system of science. In the last few decades, however, federal requirements for the regulation of research ethics have grown via rules established first by the granting councils individually and then in the late 1990s reinforced as a regulatory regime by the granting agencies collectively. Federal science-based departments such as Health Canada have also extended their reach as regulators into the researcher's domain. Moreover, universities themselves have had to operate more overt structured ethics rules and compliance systems, including appeal and review processes. Overall, therefore, this chapter draws out a further key and changing feature of federal research and innovation policy, revealed in this case through regulatory policy mechanisms embedded in rules, guidelines, and standards, but functioning as an extra governance layer built upon the earlier foundation of self-regulation and professional research norms.

The effects of these federal rules and compliance processes on Canadian universities have been significant. The adoption of the one-size-fits-all approach of the TCPS is appropriate as an additional layer of reassurance to governments and taxpayers who fund research. However, the chapter has shown that it may well be inappropriate for the natural and social sciences, and may potentially make it more difficult to get research done for some researchers.

In addition, the complexity of the ethics regime for research involving humans, which now involves international regulators as well, may lead to confusion on the part faculty, staff, and students. Not knowing which rules apply, coupled with more levels of review, may cause difficulties, increased costs, and delays for some university researchers, notably those involved in science and bio-medical research.

Also, as we have seen, the ethics rules make Canadian universities an internal regulator of faculty, staff, and students in a much more explicit and overt way. With this role, future research will be required to understand the needs of universities and help them to fulfil their duties as internal ethics regulators for the federal granting agencies. This research will have to inquire more into what the driving motivations for faculty members are to serve on ethics review mechanisms (for example, a REB). The issue of how universities can assist faculty members serving on ethics review mechanisms to manage their duties in addition to their teaching, research, and other administrative duties is also important. The question of what structural changes are necessary to assist these ethics review mechanisms in discharging their duties is a further institutional issue. In light of the likely creation of a new accreditation or alternative system for the protection of human research subjects, questions need to be raised in relation to the impact of such a system on Canadian universities. What supports do universities need in order to manage the new administrative demands resulting from the introduction of an accreditation system or alternative system for the protection of human research subjects? How will such a system impact the various disciplines and fields of research within Canadian universities? How will smaller liberal arts universities be impacted?

What is clear from the discussion in this chapter is that the federal regulatory role is more overt and is increasingly complex. Thus, any future changes must consider the needs of Canadian universities to regulate, not just manage, faculty, staff, and students in their pursuit of knowledge.

REFERENCES

Baird, Patricia. (2003). 'Getting It Right: Industry Sponsorship and Medical Research.' *Canadian Medical Association Journal* 168 (10): 1267–9.
Baker, Heather. (n.d.). 'Research Ethics Boards and Their Limitations.' Canadians for Health Research. Available from: http://www.chrcrm.org/main/modules/pageworks/index.php?id=244&page=015 (accessed 1 October 2006).
Berg, Bruce. (2007). *Qualitative Research Methods for the Social Sciences*. 6th ed. Boston: Pearson/Allyn & Bacon.
Bevan, Joan. (2002). 'Towards the Regulation of Research Ethics Boards.' *Canadian Journal of Anaesthesia* 49: 900–6.
CCAC. (1999). CCAC Assessment Panel Policy. Canadian Council on Animal Care. Available from: http://www.ccac.ca/en/CCAC_Main.htm (accessed 1 October 2006).
– (2003–04). *Annual Report*. CCAC. Available from: http://www.ccac.ca/en/CCAC_Main.htm (accessed 1 October 2006).

- (2006). Terms of Reference for Animal Care Committees. CCAC. Available from: http://www.ccac.ca/en/CCAC_Main.htm (accessed 1 October 2006).
CIHR, NSERC, and SSHRC. (1998; with 2000, 2002, and 2005 amendments). *Tri-Council Policy Statement: Ethical Conduct for Research Involving Humans.* Canadian Institutes of Health Research, Natural Sciences and Engineering Research Council of Canada, and Social Sciences and Humanities Research Council of Canada.
Canadian Medical Association Journal. (2001). 'Editorial: Look, No-Strings: Publishing Industry-Funded Research.' *Canadian Medical Association Journal* 165 (6): 733–5.
Dinsdale, Henry. (2005). 'Professional Responsibility and the Protection of Human Subjects.' *Health Law Review* 13 (2 and 3): 80–5.
- (2006). 'Accreditation and Research Involving Humans in Canada: Background and Prospects.' *NCEHR Communiqué* 14 (1): 4–7.
Ferris, Lorraine. (2002). 'Industry-Sponsored Pharmaceutical Trials and Research Ethics Boards: Are They Cloaked In Too Much Secrecy?' *Canadian Medical Association Journal* 166 (10): 1279–80.
Gauthier, Clément. (2004). 'Process for Change – Development and Implementation of Standards or Animal Care and Use in Canada.' *The Development of Science-Based Guidelines for Laboratory Animal Care: Proceedings of the November 2003 International Workshop.* Available from: http://books.nap.edu/openbook.php?isbn=0309093023 (accessed 12 January 2009).
Gibson, Elaine, Francoise Baylis, and Steven Lewis. (2002). 'Dances with the Pharmaceutical Industry.' *Canadian Medical Association Journal* 166 (4): 448–50.
Gold, Jennifer. (2003). 'Watching the Watchdogs: Negligence, Liability and Research Ethics.' *Health Law Journal* 11: 153–76.
Gold, Jennifer, Michelle Laxer, and Paula Rochon. (2003). 'Monitoring Contracts with Industry: Why Research Ethics Boards Must be Involved.' *Health Law Review* 11 (3): 13–16.
Government of Canada. (2002a). *Response to the Recommendations to Health Canada of the Coroner's Jury Investigation into the Death of Vanessa Young.* Ottawa: Health Canada.
- (2002b). *Health Products and Foods Branch Inspectorate – Inspection Strategies for Clinical Trials Policy.* Ottawa: Health Canada.
- (2003). *Food and Drug Regulations, Part C – Drugs, Division 5.* Ottawa: Health Canada.
Grayson, J. Paul, and Richard Myles. (2004). 'How Research Ethics Boards are Undermining Survey Research on Canadian University Students.' *Journal of Academic Ethics* 2: 293–314.

Halliwell, Janet. (2005). 'The National Discussion on Protection of Human Subjects in Research.' Ottawa: SSHRC.
– (n.d.). *'Ethics: A Complex and Demanding File – Diverse Players, Competing and Conflicting Interests, Complex Governance Structures.'* Ottawa: SSHRC.
Hirtle, Marie. (2003). 'The Governance of Research Involving Human Participants in Canada.' *Health Law Journal* 11: 137–52.
House of Commons Standing Committee on Health. (2004). *Opening the Medicine Cabinet: First Report on Health Aspects of Prescriptive Drugs.* Ottawa: Health Canada.
Interagency Advisory Panel on Research Ethics. (N.d) 'Introductory Tutorial for the TCPS.' Available from: http://pre.ethics.gc.ca/english/tutorial/00_intro_overview_context.cfm (accessed 16 November 2007).
International Committee of Medical Journal Editors. (2007). 'Uniform Requirements for Manuscripts Submitted to Biomedical Journals: Writing and Editing for Biomedical Publication.' Available from: http://www.icmje.org/index.html#ethic (accessed 16 November 2007).
Johnson, John, and David Altheide. (2002). 'Reflections on Professional Ethics.' In *Walking the Tightrope: Ethical Issues for Qualitative Researchers*, ed. Will van den Hoonaard, 59–69. Toronto: University of Toronto Press.
Kellner, Florence. (2002). 'Yet Another Coming Crisis? Coping with Guidelines from the Tri-Council.' In *Walking the Tightrope: Ethical Issues for Qualitative Researchers*, ed. Will van den Hoonaard, 26–33. Toronto: University of Toronto Press.
Lemmens, Trudo. (2005). 'Federal Regulation of REB Review of Clinical Trials: A Modest but Easy Step Towards an Accountable REB Review.' *Health Law Review* 13 (2/3): 39–50.
Lougheed, Tim. (2003). 'A Question of Ethics.' *University Affairs* (June/July): 10–13.
McDonald, M. (2005). 'Special Issue: Canadian Governance for Ethical Research Involving Humans. *Health Law Review* 13 (2/3): 5–12.
– (n.d.). 'What's Right, What's Missing, What's Next?' Canadians for Health Research. Online at: http://www.chrcrm.org/main/index.php? (accessed 1 October 2006).
Miller, J.R. (2006). 'A Short History of SSHRC.' Unpublished paper.
NCEHR. (2006). *Task Force for the Development of an Accreditation System for Human Research Protection Programs – Draft Final Report.* 27 January.
Rehnquist, Janet. (2001). 'The Globalization of Clinical Trials: A Growing Challenge in Protecting Human Subjects.' U.S. Department of Human and Health Services, Office of Inspector General.
Schuchman, M. (2005). *The Drug Trial: Nancy Olivieri and the Science Scandal that Rocked the Hospital for Sick Kids.* Toronto: Random House Canada.

Schuppli, Catharine, and Michael McDonald. (2005). 'Contrasting Modes of Governance for the Protection of Humans and Animals in Canada: Lessons for Reform.' *Health Law Review* 13 (2 and 3): 97–106.

Sykes, Susan. (2005). 'Protecting Human Participants: Living with Evolving Research Ethics Guidelines.' Waterloo: University of Waterloo.

Thompson, Jon, Patricia Baird, and Jocelyn Downie. (2001). 'Report of the Committee on the Case Involving Dr. Nancy Olivieri, the Hospital for Sick Children, the University of Toronto and Apotex, Inc.' From: http://www.caut.ca/en/issues/academicfreedom/OlivieriInquiryReport.pdf (accessed 1 November 2006). Reprinted as *The Olivieri Report: The Complete Text of the Report of the Independent Inquiry Commissioned by the Canadian Association of University Teachers*. Toronto: CAUT /James Lorimer, 2001.

van den Hoonaard, Will. (2002). 'Introduction: Ethical Norming and Qualitative Research.' In *Walking the Tightrope: Ethical Issues for Qualitative Researchers*, ed. Will van den Hoonaard, 3–16. Toronto: University of Toronto Press.

van den Hoonaard, Will, Lisa Givens, Joseph Levy, Michelle McGinn, Patrick O'Neill, and Ted Palys. (2004). *Giving Voice to the Spectrum: Report of the Social Sciences and Humanities Research Ethics Special Working Committee to the Interagency Advisory Panel on Research Ethics*. Ottawa: Government of Canada.

Weijer, Charles. (2001). 'Continuing Review of Research Approved by Canadian Research Ethics Boards.' *Canadian Medical Association Journal* 164 (9): 1305–6.

11 Universities and Knowledge Transfer: Powering Local Economic and Cluster Development

DAVID A. WOLFE

Previous chapters have in various ways indicated the important changed relations and interactions between federal research and innovation policy and universities. These relationships have been shown to be both direct and indirect and involve complex uses of spending, regulation, and changed boundary organizations and institutions. In this chapter, the federal role is very indirect in part because this chapter focuses on universities, knowledge transfer, and regional economic development and cluster development, realms of research and innovation that are quite simply spatially more distant from the federal government and also decidedly non-linear in nature. Examples will certainly emerge in this chapter of the involvement and presence of federal labs and programs in regional development and local-regional cluster formation. However, these kinds of federal involvement tend to be embedded within much more complex and highly dynamic networks of research and learning involving the private sector, provincial governments, local governments, local universities, and local communities.

In the transformation of the industrial economies towards more knowledge-based ones, universities have emerged as central actors in the process of regional economic development and cluster formation. No longer limited to their traditional roles of teaching and conducting research, the widely cited examples of Stanford in the growth of Silicon Valley and MIT in the development of Route 128 imply that universities have become key drivers of innovation and 'major agents of economic growth.' Consequently, local and regional policymakers have come to view the presence of a research university as a critical asset for their local economies, with untapped reservoirs of commercializable knowledge, waiting to be taken up by firms and applied. While the presence of a leading research university in a community is a critical asset, it is not sufficient in itself to stimulate strong regional economic growth (Wolfe 2005). Current research on the broader contribution of universities confirms that they do play an essential role in the process of cluster formation and regional economic development, but we need a more contextualized understanding of the specific way in which university-based research influences the local economy.

Recent research by members of the Innovation Systems Research Network (ISRN) on the formation and growth of industrial clusters across Canada suggests that universities tend to be followers of technological innovation rather than leaders – 'catalysts' rather than 'drivers' (Doutriaux 2003; Wolfe and Gertler 2004; Wolfe 2003; Wolfe and Lucas 2004; Wolfe and Lucas 2005). This chapter reports on the findings of that study with respect to the role of universities in cluster formation. It critiques the mechanistic assumptions that are often made about universities as generators of commercializable knowledge, and argues that the task of transferring knowledge from universities to industry is far more complex, and the role of universities in local economies far more varied, than linear conceptions of the innovation process assume. As Mowery et al. have recently argued, 'Any assessment of the economic role of universities must recognize the numerous, diverse channels through which university research influences industrial innovation and vice versa' (2004, 179).

The Relationship between Research and Innovation

The shift to a more knowledge-based economy embodies a number of changes in both the production and application of new scientific knowledge, with critical implications for the processes of knowledge transfer and regional economic development. One of the most significant of these changes involves the relation between the codified and tacit dimensions of knowledge. Following Michael Polanyi (1967), tacit knowledge refers to knowledge or insights acquired by individuals in the course of their scientific work that is ill-defined or uncodified and that they themselves cannot fully articulate. The development of such internalized or 'personal knowledge' often requires an extensive learning process. It is based on skills accumulated through personal experience and expertise. It also emphasizes the learning properties of individuals and organizations; individuals or groups working together for the same firm or organization often develop a common base of tacit knowledge in the course of their research and production activities (Nelson and Winter 1982, 76–82).

The second change concerns the centrality of learning for the innovation process. Lundvall argues that the knowledge frontier is moving so rapidly that access to, or control over, knowledge assets affords merely a fleeting competitive advantage. It may be more appropriate to describe the emerging paradigm as that of a 'learning economy,' rather than a 'knowledge-based' one (2004). Recent work indicates that innovation involves a capacity for localized learning within firms, between firms, and between firms and supporting institutions. Learning in this sense refers to the building of new competencies and the acquisition of new skills, not just gaining access to information or codified scientific knowledge. In tandem with this development, forms of knowledge that cannot be codified and transmitted electronically (tacit knowledge) increase in value, along with the ability to acquire and assess both codified and tacit forms of knowledge, in other words, the capacity for learning. The tacit dimension of knowledge is particularly significant for regions and communities, for tacit knowledge tends to be embedded in local institutions and networks of firms (Morgan and Nauwelaers 1999, 6). The ability of firms within a regional economy to find and appropriate tacit knowledge produced by individual workers or communities of knowledge often draws upon a common culture that facilitates the flow of knowledge among them and is supported by a common set of regional institutions (Gertler 2002, 2003).

A more accurate appreciation of the role played by universities in local and regional innovation systems requires an understanding of the institutional and interpersonal linkages between universities and firms and how those linkages contribute to knowledge transfers between the two. Keith Pavitt stresses that the tacit component of scientific and technological knowledge means that the most important, and commercially valuable, knowledge is often embodied in the personal knowledge, skills, and practices of individual researchers (1991). He builds on Nathan Rosenberg's claim that to assimilate and benefit from external research, firms have to develop a considerable capacity for research themselves (Rosenberg 1990). Pavitt maintains that knowledge transfers are mainly person-embodied and that policies that focus narrowly on directing the results of basic research towards specific goals or targets ignore the considerable indirect benefits across a broad range of scientific fields that result from the training of researchers and unplanned discoveries. Of crucial importance are the role of skills, the networks of researchers, and the development of new capabilities on the part of actors and institutions in the innovation system.

A key implication of this argument is that firms require a strong contingent of highly qualified research scientists and engineers as a prerequisite to the ability to absorb and assess scientific results, most frequently recruited from institutions of higher education. The members of this scientific and engineering labour force bring with them not only the knowledge base and research skills acquired in their university training, but often, more importantly, a network of academic contacts acquired during their university training. This underlines Pavitt's point about the person-embodied nature of knowledge transfer. A study by Wendy Faulkner and Jacqueline Senker explores the relationship from the perspective of the innovating organization, focusing on its knowledge requirements and trying to develop a better understanding of the knowledge flows from academia to industry. The researchers probed for links between the firms and universities and the types of knowledge flowing to the firms; they also attempted to determine the degree of formality of these links and the relative importance of tacit versus codified knowledge. They conclude that partnering with universities contributes most to firm innovation through an exchange of tacit knowledge and that the channels for communicating this knowledge are often informal. Such informal linkages are both a precursor and a successor to formal linkages and many useful exchanges of research materials or access to equipment take place through non-contractual barter arrangements (Faulkner and Senker 1995).

Universities in the Knowledge-Based Economy

As previous chapters have shown, the expanded research role that universities have come to play in the knowledge-based economy is very much the result of the post-war 'social contract' for science that was forged in the aftermath of World War II. Based on the success of the wartime research efforts in mobilizing national scientific research capabilities in aid of the war effort, the social contract saw governments in Canada and the other industrial countries willing to fund massive investments in basic research in the expectation of long-term economic benefits, while leaving the principal research institutions, the universities, autonomous in the conduct of that research (Martin 2003; Brooks 1986).

However, the essential elements of the social contract for science have been subject to increasing strain in the past two decades. The changes that have impacted on the university system are part of the broader trends in the shifting location for the performance of basic and applied R&D. At issue is the changing nature of the relationship between the universities and the broader innovation system in which they are embedded, as well as the process of scientific investigation and discovery that underlies the knowledge production function (Gibbons et al. 1994). Since the early 1980s, private firms have expanded their research linkages with universities, partly in response to the rising cost of conducting R&D. Under competitive pressure to introduce new products, processes, and services more quickly, many large corporations have restructured their R&D operations to link research programs more tightly with product development processes. Broader-based inquiries into fundamental science have consequently been scaled back in many firms.

Universities, in turn, have come under increasing pressure to supplement their traditional role in the conduct of basic research with more applied research activities, frequently based on university–industry partnerships. Expectations about the changing role for universities have been driven by three intersecting trends: 1) the linking of government funding for academic research and economic policy; 2) the development of more long-term relationships between firms and academic researchers; and 3) the increasing direct participation of universities in commercializing research (Etzkowitz and Webster 1998; Geiger 2004).

As previous chapters have shown, these trends have resulted in the proliferation of a new range of university–industry technology transfer mechanisms, including: industry liaison offices (ILOs) and technology transfer offices (TTOs) in universities, research parks affiliated with universities, university–industry consortia, research institutes and centres of excellence, regional development organizations, and spin-off firms. These mechanisms perform a wide range of functions, including the negotiation of industrial research contracts, the identification of opportunities for university research in the marketplace, and the facilitation of licensing or patenting of research results or the spin-off of new firms.

However, the mere proliferation of these activities should not be equated with an increase in their effectiveness or efficiency (Doutriaux and Barker 1995; Etzkowitz 1999). In part, this shift reflects the change in the nature of business R&D described above, but it is also the result of a parallel expectation on the part of government that their investments in basic research will produce an increased economic return.

While linkages between universities and industry have proliferated in the past decade and a half, the resulting contribution to regional economic development has not always materialized in the anticipated ways. The reason for this is that university research is part of a much larger complex of factors that bear upon the innovation process and, as noted above, the flows of knowledge that underpin this process are far from unidirectional. As Fumio Kodama and Lewis Branscomb argue,

> disappointment awaits those who expect quick results from university-based high-technology strategies for industrial renewal. First-rank research universities can and most often do make a large and positive contribution to economic performance, regionally and nationally. But to understand the effects we should not focus on the style and content of the transactions with firms but rather look at the university as a pivotal part of a network of people and institutions who possess high skills, imagination, the incentive to take risks, the ability to form other networks to accomplish their dreams. (1999, 16)

Universities and Regional Economic Development

The proximity effect of knowledge transfer provides a strong clue as to why universities are increasingly seen as an essential element to the process of regional economic development and for stimulating the formation of clusters, especially in knowledge-intensive industries, such as information and communications technology or biotechnology. Knowledge transfers between universities and industrial firms are iterative, highly personalized, and, as a consequence, often highly localized. This underscores the significance of geographical proximity for the process of knowledge transfer. Proximity to the source of the research is important in influencing the success with which knowledge generated in the research laboratory is transferred to firms for commercial exploitation, or process innovations are adopted and diffused across developers and users. In addition to generating new knowledge through the conduct of basic research, universities provide both formal and informal technical support, as well as specialized expertise and facilities for ongoing, firm-based R&D activities.

As national and regional innovation systems become more interconnected and as the knowledge base required to support the production of 'complex technologies' becomes more highly diffused, university research also becomes important to local firms not just for the transfer of knowledge generated through its own research activities, but also as a conduit enabling firms to access knowledge from the 'global pipelines' of international academic research networks (Bathelt et al. 2004; OECD 1999). Increasingly, they also serve as attractors of talent from elsewhere that contributes to the 'thickness' of the local labour market (Florida 2002; Betts and Lee 2005). Finally, rather than acting as 'ivory towers' insulated from their community, they often function as 'good community players' that facilitate local linkages and networks (Wolfe 2005; Betts and Lee 2005).

Proximity to the source of the research is important in influencing the success with which knowledge generated in the research laboratory is transferred to firms for commercial exploitation, or process innovations are adopted and diffused across researchers and users. According to Audretsch, 'the theory of knowledge spillovers ... suggests that the propensity for innovative activity to cluster spatially will be the greatest in industries where tacit knowledge plays an important role' (2002, 170). Recent research indicates that small, and often new, firms which conduct little basic research themselves are able to grow by taking advantage of the research activities in proximate universities (Audretsch 2002, 168). But the most dynamic local universities are those which actively engaged with the broader process of social and economic development in their local community (Walshok 1995). A comprehensive study of universities and the development of regional industry clusters in the U.S. concluded that this is most successful when the knowledge assets of the university are effectively aligned with the diverse needs of local firms:

> A large base of research and development is required but not sufficient. The university must also address the business, workforce, and community issues. The university must be aligned with regional interests and industry clusters across a broad spectrum, not just in terms of technical knowledge. (Paytas et al. 2004, 34)

The role played by universities in attracting and retaining talent has also been identified as a critical contribution to the growth of dynamic urban economies. Highly educated, talented labour flows to those places that have a 'buzz' about them – the places where the most interesting work in the field is currently being done. Locations with large talent pools reduce the costs of search and recruitment of talent; they are also attractive to individuals who are relocating because they provide some guarantee of successive job opportunities. Inbound talented labour represents knowledge in its embodied form flowing into the region. Such flows act to reinforce and accentuate the knowledge assets already assembled in a region. Knowledge diffusion occurs more rapidly among co-located firms than widely dispersed ones due to the circulation of highly skilled individuals within the local labour market. Universities can play a central role in the development of a local talent pool. Gertler and Vinodrai characterize universities as 'anchors of creativity' in producing, attracting, and retaining highly skilled talent, while they simultaneously create an open and tolerant attitude in the communities in which they are located. This in turn reinforces the conditions needed to attract and retain talent in the local community (2005).

Universities emerge from this analysis as multi-faceted economic actors that are embedded in regions and participate actively as important institutional actors in building and sustaining local networks and flows of knowledge, and in linking them with global ones. In general this body of research has found that research-intensive universities often play a central role in regional economic development, but the precise nature of their contribution is varied and complex. Rosenberg argues that American commercial success in high-technology sectors of the economy 'owes an enormous debt to the entrepreneurial activities of American universities' (2003, 116), but cautions at the same time that the precise impact of the university varies significantly across industrial sectors and regional contexts. The Innovation U. project, a recent study of how a small group of research universities in the U.S. are using their technological strength to build links with industry, identifies the emergence of a new twenty-first century model of an 'entrepreneurial research university' that 'aggressively partners with technology-based industry and regional economic development interests, exhibits and encourages entrepreneurial behavior, and champions these new directions in its public pronouncements and internal values' (Tornatzky et al. 2002).

Universities as Knowledge Poles for Cluster Development

In an earlier paper Gertler and I examined the nature of path dependency in the process of cluster formation and, in particular, the role that public policy, including support for post-secondary research institutions, played in that process (Wolfe and Gertler 2006). In many of the cases that comprised the ISRN study of industrial clusters, governments played an important role in assisting the formation of regional clusters, although that role varies considerably across the regions. Public policies that create a strong knowledge base in the regional economy and contribute to the creation of a well-educated workforce establish the local antecedents that can support the emergence of clusters and contribute to local economic development. While a strong research infrastructure and a thick labour market are distinctly local phenomena, in most industrial countries they are not exclusively the result of local, or even state and provincial government policies; the presence of the senior level of government lurks in the background. In regions with a particularly strong and diverse industrial base, such as Waterloo, the cluster benefited indirectly through government support for the local universities. Although the University of Waterloo can be considered an anchor institution because of its key role in the creation of local talent and spin-off firms, the university had little direct involvement in the innovative activities of the high-tech cluster. Local firms played the key role in mobilizing the resources needed to build the cluster, yet firms overwhelmingly attribute the dynamism of the cluster and the economic success of the region to the presence of the university (Bramwell and Wolfe 2008).

And while universities and public research institutes can play a supporting role in the development of the cluster, as was the case with the presence of federal laboratories in the national capital in attracting the Bell Northern Research Laboratories to Ottawa, their role is far less direct or instrumental than is often presumed. A critical factor that seeds the growth of a cluster is a deep pool of highly skilled labour, or a unique mix of skill assets, often produced by the post-secondary educational institutions. The contribution of universities to local cluster development should not be viewed simply as a source of scientific ideas for generating new technology to transfer to private firms, or as a source of new firm formation as research scientists spin findings out of their laboratories into new start-ups. Although successful research universities perform these functions, they play a more fundamental role as providers and attractors of talent to the local and regional economy and as a source of civic leadership for the local community. Drawing upon a subset of the ISRN case studies, Doutriaux found no direct causal relationship between the presence of a university and local high technology development in eleven high technology clusters (2003). His observation is confirmed in a broader cross-section of the case studies which found that universities often act as catalysts for cluster development by training local talent and contributing to the local knowledge base, but they were rarely the key drivers of cluster formation in themselves. Many of the leading research-intensive universities in Canada have proven themselves highly effective at responding to 'market' signals and expanding their research and teaching activities in fields that are most heavily in demand with local industry.

Tangible links between the university and industry, be they in the form of research collaborations, co-op programs, or more general education, build a region's innovative capacity by acting as two-way conduits of knowledge that keep firms at the leading edge of innovation and keep universities relevant to local industry. A key observation that emerges from the case studies, however, is the way in which the role of the university in the region changes over time. Universities act as sources of new invention and innovation, performers of R&D that may be taken up and applied by firms in a local cluster, and as generators of human capital; but as local clusters mature through the various stages of the cluster life cycle, they also play an important role in supporting the incremental innovation that keeps firms competitive. As both firms in a cluster mature, and their products mature, they rely upon universities and public research institutes, less as sources of new knowledge, and more for their expertise in providing technical solutions to challenges in developing next-generation hardware and software products. This role also varies considerably across the technology base of the different industrial clusters. Relatively few of the clusters in information and communications technology (ICTs), or in the more traditional manufacturing industry, owe their location in a specific region to the presence of a research university. This is clearly not the case in the most science-based clusters, such as biotechnology, where they have mostly been anchored by the research output of either local universities or the presence of a National Research Council Institute.

The research results on the ICT clusters in the ISRN project documents the important contribution made by the research infrastructure – both public research laboratories and post-secondary educational institutions (Bramwell et al. 2008; Lucas, Sands, and Wolfe 2009; Chamberlin and de la Mothe 2003; Davis and Schaefer 2003; Langford et al. 2003; Kéroack et al. 2004). The findings underline the fact that direct seeding of the cluster by post-secondary institutions is the exception, rather than the rule. The case studies indicate that universities and research institutes act primarily as attractors of inward investments by lead anchor firms interested in tapping into the knowledge base of the local community, or its local buzz, and as providers of the talent pool that firms in the cluster draw upon, rather than as direct initiators of cluster development. In this respect, universities also act as part of the network linking actors in the local cluster to the global pipelines that are essential to sustain critical knowledge flows for local firms in the cluster. Successful research universities also attract leading scientists, further reinforcing their linkages to external knowledge flows through the extensive network of contacts they bring to their new location.

A case that illustrates the inadvertent role that other public sector research institutions can play is provided in the case of the telecommunications and photonics cluster in Ottawa, which originated partly with the judicial decision in the U.S. to force the Western Electric Company to divest itself of its subsidiary, the Northern Electrical Manufacturing Company (now Nortel), in the late 1950s. Cut off from its sources of innovation and research, Northern Electric searched for a location to establish its own facility. It eventually bought a substantial tract of land on the outskirts of Ottawa to be the home of Bell Northern Research, largely because it viewed the presence of the National Research Council laboratories and the Communications Research Centre in the nation's capital as a substantial draw for the highly skilled research scientists and engineers it expected to populate its research facility. Many leading entrepreneurs in the Ottawa telecommunications and photonics cluster began their careers as researchers for BNR or another subsidiary, Microsystems International. However, the research universities in the city, as well as the city's community college, have responded in substantive fashion to the growth of the cluster by contributing to the development of local firms through the expansion of their teaching and research activities. The universities have also worked closely with the National Research Council in the development of new facilities, such as the Photonics Fabrication Centre, to support the further growth of local firms in the cluster.

> The Ottawa story emphasizes the critical importance in cluster development of deeply rooted R&D strength. It also clearly underscores the fact that access to technology demands the presence of world-class scientific research institutions. Only through the combined impact of public and private sector research activity could Ottawa have spawned its own homegrown high-tech industry. (Mallet 2002, 6)

In a series of interviews conducted for the case studies, universities were identified as a source of strength for the ICT sector, both in terms of their ability to provide a steady stream of highly skilled personnel and as a strong base of research with close links to industry. Industry representatives feel that specific programs such as the co-op programs at Waterloo and other universities have been effective at moving students into industry settings. In addition to providing a strong talent base for firms located in the clusters to draw upon, the university research infrastructure is important for the clusters in two additional respects – first, as a key source of new ideas for domestic companies, both in terms of spin-offs and knowledge transfer; and second, as a factor contributing to the reputation of the key clusters – in Ottawa, the Greater Toronto Area, and Waterloo – thus helping to attract large foreign firms to invest in the province. The case of Cisco is widely cited as the most significant inward investment to the Ottawa cluster, but Alcatel, Lucent, and, most recently, Google in Waterloo, have also been mentioned. IBM, with one of its Centres for Advanced Studies located in its software laboratories in Markham, just north of Toronto, enjoys a strong working relationship with the University of Toronto and has expanded its presence in the Ottawa cluster as well. Through the acquisition of two local software companies – Tarian Software and Rational Software Corporation – it tripled the size of its Ottawa laboratories. This expansion was furthered with the opening of a new Centre for Advanced Studies in the Ottawa laboratories to give graduate students from Carleton and the University of Ottawa experience in working with experienced software engineers in the development of new programs for database management (Pilieci 2003).

Many companies are expanding their investments in the university research base – through direct funding of basic research, affiliation with federal and provincial Centres of Excellence, or partnering on more applied research initiatives. Companies cite the positive benefits that have flowed from recent federal and provincial increases in university funding through programs such as the Centres of Excellence, the Canada Foundation for Innovation, or the Ontario Research Fund. The two largest players in the telecommunications sector, Bell Canada and Nortel Networks, both launched major university-based research initiatives in the late 1990s. In 2000, Nortel was funding $15 million of research at eleven Ontario universities and had invested an additional $18 million to create two dedicated institutes, the Nortel Institute of Optical Electronics at the University of Toronto and the Software Institute at the University of Waterloo. For its part, Bell Canada invested $35 million over three years in its Bell

University Laboratories program at the Universities of Toronto and Waterloo and subsequently expanded the program across the country. Although the level of corporate research funding to universities declined in the recession of the early 2000s, the strong links to the research base remain intact.

The University of Waterloo (UW) has exerted a profound impact on the development of high-tech industry and the shape of the regional economy. Several strategic decisions made during its formative years laid the groundwork of expertise, research capacity, and talent that characterizes its role as a catalyst for the region's current high-tech economy. All accounts of the origins of this cluster link its roots to the far-sighted vision of a key group of business leaders to create a new university in the region in the late 1950s, in a period when the provincial government (with financial support from the federal government) was expanding the post-secondary education system.

Even more influential were subsequent decisions to focus the core strengths of the university in the sciences and engineering and to establish what has become one of the most successful co-op education programs in North America. The founders of many of the firms that populate this cluster are graduates of the university, and many started their companies with core technologies developed while they were at the university. Even the most internationally successful of these firms maintain their primary research base in the Waterloo area because of their ability to draw upon a highly trained group of science and engineering graduates from the university (Nelles et al. 2005). Representatives of the business and industry association sectors cite the strong entrepreneurial culture at the University of Waterloo and the encouragement that faculty receive to develop and exploit their innovations as a critical factor in the growth of the ICT cluster in Canada's Technology Triangle. They agree emphatically on the driving role that the university's research base has played in the recent growth and expansion of the ICT cluster in their region. The region is currently reaping the benefits of investments in the post-secondary research and education base made in the late 1950s and 1960s.

Much of the University of Waterloo's success at linking with both local and non-local industry is largely attributable to four well-known characteristics: the provision of R&D support to local firms; the attraction, retention, and training of top calibre researchers; the active facilitation of entrepreneurial activities; and the interactive exchange of tacit knowledge. From its inception, the university developed and has maintained a strong international reputation for academic excellence in science, math, and engineering. An innovative Intellectual Property (IP) Policy, where full ownership of IP rests with the creator, allowing the individual faculty or student to commercialize their ideas, has been credited with the large number of high-profile start-ups and spin-offs in the region. Equally influential is its cooperative education program, where students complete work terms in industry as part of their curriculum. This training innovation was adopted in the early days of the institution and remains the most successful of its kind in Canada. Many people credit Waterloo's success to its academic and research excellence coupled with these last two innovations (Bramwell and Wolfe 2008).

This characterization goes part way in explaining the significance of the university for the regional economy, but does not capture the full range and depth of linkages with local industry. The University of Waterloo demonstrates a multifaceted capacity for knowledge transfer to the local economy. In terms of knowledge creation, UW provides technical support for ongoing firm-based R&D activities through project-oriented consulting and joint research projects. In terms of human capital creation, through its co-operative education program and graduate training, the university generates a large pool of highly qualified and experienced scientists and researchers, who are attuned to the research and technology needs of industry, and also attracts scientific talent from elsewhere to the local community. In terms of global linkages, the local knowledge transfers also draw upon the university's linkages with 'global pipelines' of new knowledge through the involvement of faculty with international research networks. Finally, the university acts as an engaged entrepreneurial institution – or 'good community player' – that is embedded in the local economy and shapes and supports the local networks and flows of knowledge that underpin a highly successful 'entrepreneurial' culture. In summary, UW emerges as multi-faceted actor in the Waterloo region that actively participates in building and sustaining local networks and flows of knowledge, and in linking them with global ones (Bramwell and Wolfe 2008).

The presence of strong regional research institutions also provides highly specialized consultants and attracts highly talented workers to a region. One of the strengths of the emerging photonics cluster in Quebec City is the strong links between public research institutions and local firms. This has expanded into strong networking links throughout the industry and region. One of the positive developments in the New Brunswick cluster is the recent initiative by the National Research

Council to create a local research centre in ICT affiliated with the local university. This will help create a local knowledge base and, most important, increase the R&D linkages between the university and local firms (Kerouack et al. 2004; Davis and Schaefer 2003).

Universities play a fundamentally more central role in generating the basic research that feeds the growth of emerging clusters in the life sciences (Gartler and Vinodrai 2009). Given the importance of R&D to innovation in the life sciences, it is not surprising that the presence of a well-developed local R&D system, comprised of a mix of government laboratories, universities, and technology transfer offices, has been a common institutional feature for most of the case studies of biotechnology clusters across the country, regardless of whether the region studied was a diverse 'megacentre' or a smaller, specialized life science cluster. The mere presence of a public research institution was not always the catalyzing factor in cluster development; however, these research institutions were critical to anchoring life science activity to particular regions, especially because of their relationships with local firms and their role in generating talent for the local labour market.

In the cases of Montreal, Vancouver, and Saskatoon, institutional leaders such as specialized centres, institutes, and hospitals with strong research foci often coordinated local research activities. For example, in Vancouver, a range of regional and local actors collectively contribute to Vancouver's expertise in human health. These include the Centre for Integrated Genomics Canada (CIG), Genome BC, BC Cancer Research Centre, Canadian Genetic Diseases Network (NCE), Canadian HIV Trials Network, Michael Smith Foundation for Health Research, SARS Accelerated Vaccine Initiative, and the Vancouver Coastal Health Research Institute. Furthermore, UBC's efforts through the UILO resulted in a number of spin-off firms. In Saskatoon the NRC-PBI, AAFC Saskatoon Research Centre, POS Pilot Plant, and the University of Saskatchewan have agricultural-related research strengths and act to coordinate research efforts with local private sector firms. And in Montreal, the publicly funded NRC-BRI and its research universities (e.g., McGill University) co-evolved with nascent biopharmaceutical private sector efforts. These public research institutions continue to work in what is considered by outsiders as a 'tripartite symbiotic relationship' with the provincial government and private sector, and in this way help to shape the direction of research.

In other cases, high calibre public research centres and local universities have not only produced new research and knowledge, but have also acted as magnets to attract 'star scientists' and other highly skilled workers to the region. For example, Toronto's downtown is the central location for a number of research-intensive life science institutions. The newly created MaRS Centre located in Toronto's Discovery District is designed to be a convergence innovation centre, providing opportunities to bring biomedical and life science researchers in contact with other diverse fields of knowledge in art, science, and technology. The centre also allows for the co-location of professional service firms, technology transfer offices, research and community networking organizations, and small, mid-size, and large companies. This concentration of public research institutions has collectively attracted a number of top scientists and researchers, as well as the interest of venture capitalists and local start-up firms.

In the traditional manufacturing and resource case studies covered by the ISRN research, the interface between industrial firms and educational institutions involves both knowledge transfers and people flows (Warrian and Mulhern 2009). This relationship is critical for the ability of these industries to adapt to the knowledge economy. For executives, the people flows created by formal and informal arrangements with universities, colleges, and apprenticeship programs are the most important benefit from firm-educational linkages. There is also evidence that the core scientific skills of professors in research labs have been of significant assistance in the introduction of new processes to plants and even in problem-solving on the ground. People and knowledge flows have also generated large numbers of start-up companies on the model of the University of Waterloo. This being said, the community college system also contributes to local talent pools.

The emerging mining supply and services cluster in Sudbury draws heavily upon the growing supply of highly qualified personnel from local colleges, universities, and research centres (Robinson 2006). At the same time, public research institutions, principally Laurentian University, have moved toward becoming a major research supplier in the mining area. There is a significant transfer of research to private sector mining operations, but limited transmission to the low- and medium-tech SMEs. Furthermore, as the linkages with research groups grow in Sudbury, these increasingly involve inter-firm links. In Sudbury the universities and government laboratories have played a secondary role historically, but appear to be increasing in importance.

Despite the historical weakness of the Ontario auto parts and assembly industries as a source of R&D, the University of Windsor has moved dramatically over the last ten years to foster collaborative R&D with the Big Three and establish Windsor as the 'Automotive Intellectual Capital of Canada.' The university has created several research centres and Industrial Research Chairs linked to automotive engineering over the last decade. However, these research efforts are geared more towards the OEMs, and plant managers in both the 'core' automotive parts sector and the MTDM industry in Windsor made limited reference to the role played by the University of Windsor in the automotive parts cluster. On the other hand, they noted the value of the pool of well-trained skilled production workers produced by St Clair Community College and, in the past, area technical high schools (Fitzgibbon et al. 2004; Holmes et al. 2005).

Conclusions

This chapter has explored the changing role of universities in knowledge transfer and in regional development and cluster formation. In relation to the central focus of this book on changing federal–university relations in the research and innovation policy realms, it has been stressed from the outset that the federal role in this chapter's area of focus is typically very indirect. Not surprisingly, this is due to the fact that regional economic development and cluster development are realms of research and innovation that are spatially more distant from the federal government and also decidedly non-linear in nature. Examples have been given in this chapter of the involvement and presence of federal labs and programs in regional development and local-regional cluster formation, but they tend to be embedded within much more complex and highly dynamic networks of research and learning that include the private sector, provincial governments, local governments, local universities, and local communities.

Overall, the analysis has shown that the strength and vitality of universities remains essential for growth in the knowledge-based economy. Universities perform vital functions both as generators of new knowledge through their leading-edge research activities and as trainers of highly qualified labour. As most research universities will attest, the two functions are integrally linked, and when they are most effective, they contribute strongly to regional economic growth and cluster development. They also serve as magnets for investments by leading or anchor firms, drawing them into the cluster to gain more effective access to the knowledge base and local buzz.

In some instances, successful research efforts can expand the cluster by spinning off research results into new products and firms, but it is a mistake to view this as their primary purpose. The economic returns to investments in university research cannot be judged solely on the success with which research findings are transformed into commercial products. Universities must also be a vital part of the local 'economic community' by building the region's social capital and taking a leadership role in activities designed to enhance the region's economic development. Continued public support, federal and provincial, for both the teaching and research mandates of the university is essential if they are to succeed in these roles and contribute to the growth of their local and regional economies.

At the same time, communities located around the research institution cannot simply rely upon the presence of a leading research university as the 'engine of innovation' that will drive economic growth in their region. They must display both the capacity to absorb and utilize the knowledge and the skilled labour produced by the institution – in other words, a 'regional absorptive capacity' (Mallet 2002, 605) – and the social cohesion to build an economic community around their research infrastructure.

Ultimately, the most valuable contribution that universities make to this process is as providers of high skilled labour, or talent. If knowledge is rapidly becoming the central factor of production in the emerging economy, the ability to absorb that knowledge, or to learn, is the most essential skill. Learning processes are eminently person-embodied in the form of talent. This means that the role of public policy through varied policy instruments in seeding cluster development is critical. On balance, however, the public interventions that have the most effect in seeding the growth of clusters are those that strengthen the research infrastructure of a region or locality and contribute to the expansion of its talent base of skilled knowledge workers.

REFERENCES

Audretsch, David B. (2002). 'The Innovative Advantage of U.S. Cities.' *European Planning Studies* 10: 165–76.
Bathelt, Harald, Anders Malmberg, and Peter Maskell. (2004). 'Clusters and Knowledge: Local Buzz, Global Pipelines and the Process of Knowledge Creation.' *Progress in Human Geography* 28 (1): 31–56.

Betts, Julian, and Carolyn Lee. (2005). 'Universities as Drivers of Regional and National Innovation: An Assessment of the Linkages from Universities to Innovation and Economic Growth.' In *Higher Education in Canada*, ed. Charles M. Beach, Robin W. Boadway, and R. Marvin McInnis, 113–60. Montreal: McGill-Queens University Press.

Bramwell, Allison, and David A. Wolfe. (2008). 'Universities and Regional Economic Development: The Entrepreneurial University of Waterloo.' *Research Policy* 37 (8) (September): 1175–87.

Bramwell, Allison, Jen Nelles, and David A. Wolfe. (2008). 'Knowledge, Innovation and Institutions: Global and Local Dimensions of the ICT Cluster in Waterloo Canada.' *Regional Studies* 42 (1) (February): 1–16.

Brooks, Harvey. (1986). 'National Science Policy and Technological Innovation.' In *The Positive Sum Strategy: Harnessing Technology for Economic Growth*, ed. Ralph Landau and Nathan Rosenberg, 33–51. Washington, D.C.: National Academies Press.

Chamberlin, Tyler, and John de la Mothe. (2003). 'Northern Light: Ottawa's Technology Cluster.' In *Clusters Old and New: The Transition to a Knowledge Economy in Canada's Regions*, ed. David A. Wolfe, 213–34. Montreal: McGill-Queen's University Press for the School of Policy Studies, Queen's University.

Davis, Charles H., and Norbert V. Schaefer. (2003). 'Development Dynamics of a Startup Innovation Cluster: The ICT Sector in New Brunswick.' In *Clusters Old and New: The Transition to a Knowledge Economy in Canada's Regions*, ed. David A. Wolfe, 121–60. Montreal: McGill-Queen's University Press for the School of Policy Studies, Queen's University.

Doutriaux, Jerome. (2003). 'University–Industry Linkages and the Development of Knowledge Clusters in Canada.' *Local Economy* 18 (1): 63–79.

Doutriaux, Jerome, and Margaret Barker. (1995). *The University-Industry Relationship in Science and Technology*. Occasional Paper Number 11. Ottawa: Industry Canada.

Etzkowitz, Henry. (1999). 'Bridging the Gap: The Evolution of University-Industry Links in the United States.' In *Industrializing Knowledge: University-Industry Linkages in Japan and the United States*, ed. Lewis M. Branscomb, Fumio Kodama, and Richard Florida, 203–33. Cambridge, Ma.: MIT Press.

Etzkowitz, Henry, and Andrew Webster. (1998). 'Entrepreneurial Science: The Second Academic Revolution.' In *Capitalizing Knowledge: New Intersections in Industry and Academia*, ed. Henry Etzkowitz, Andrew Webster, and Peter Healey, 21–46. New York: SUNY Press.

Faulkner, Wendy, and Jacqueline Senker. (1995). *Knowledge Frontiers: Public Research and Industrial Innovation in Biotechnology, Engineering Ceramics and Parallel Computing*. Gloucestershire, U.K.: Clarendon Press.

Fitzgibbon, S., J. Holmes, T. Rutherford, and P. Kumar. (2004). 'Shifting Gears: Restructuring and Innovation in the Ontario Automotive Parts Industry.' In *Clusters in a Cold Climate: Innovation Dynamics in a Diverse Economy*, ed. D.A. Wolfe and M. Lucas, 11–41. Montreal: McGill-Queen's University Press for the School of Policy Studies, Queen's University.

Florida, Richard. (2002). *The Rise of the Creative Class: And How It's Transforming Work, Leisure, Community and Everyday Life*. New York: Basic Books.

Geiger, Roger L. (2004). *Knowledge and Money: Research Universities and the Paradox of the Marketplace*. Palo Alto, Ca.: Stanford University Press.

Gertler, Meric S. (2002). 'Technology, Culture and Social Learning: Regional and National Institutions of Governance.' In *Innovation and Social Learning: Institutional Adaptation in an Era of Technological Change*, International Political Economy Series, ed. Meric S. Gertler and David A. Wolfe, 111–34. Houndsmills and New York: Palgrave Macmillan.

– (2003). 'Tacit Knowledge and the Economic Geography of Context, or The Undefinable Tacitness of Being (There).' *Journal of Economic Geography* 3: 75–99.

Gertler, Meric S., and Tara Vinodrai. (2005). 'Anchors of Creativity: How Do Public Universities Create Competitive and Cohesive Communities?' In *Taking Public Universities Seriously*, ed. Frank Iacobucci and Carolyn Tuohy, 293–315. Toronto: University of Toronto Press.

– 'Life Sciences and Regional Innovation: One Path or Many?' *European Planning Studies* 17 (2) (February): 235–61.

Gibbons, Michael, Camille Limoges, Helga Nowotny et al. (1994). *The New Production of Knowledge: The Dynamics of Science and Research in Contemporary Societies*. London: Sage Publications.

Holmes, J., T. Rutherford, and S. Fitzgibbon. (2005). 'Innovation in the Automotive Tool, Die and Mould Industry: A Case Study of the Windsor-Essex Region.' In *Global Networks and Local Linkages: The Paradox of Cluster Development in an Open Economy*, ed. D.A. Wolfe and M. Lucas, 119–53. Montreal: McGill-Queen's University Press for the School of Policy Studies, Queen's University.

Kéroack, Mélanie, Mathieu Ouimet, and Réjean Landry. (2004). 'Networking and Innovation in the Quebec Optics/Photonics Cluster.' In *Clusters in a Cold Climate: Innovation Dynamics in a Diverse Economy*, ed. David A. Wolfe and Mathew Lucas, 113–37. Montreal: McGill-Queen's University Press for the School of Policy Studies, Queen's University.

Kodama, Fumio, and Lewis M. Branscomb. (1999). 'University Research as an Engine for Growth: How Realistic is the Vision?' In *Industrializing Knowledge: University-Industry Linkages in Japan and the United States*, ed.

Lewis M. Branscomb, Fumio Kodama, and Richard Florida, 3–19. Cambridge, Ma.: MIT Press.

Langford, Cooper H., Jaime R. Wood, and Terry Ross. (2003). 'The Origins of the Calgary Wireless Cluster.' In *Clusters Old and New: The Transition to a Knowledge Economy in Canada's Regions*, ed. David A. Wolfe, 161–85. Montreal: McGill-Queen's University Press for the School of Policy Studies, Queen's University.

Lucas, Matthew, Anita Sands, and David Wolfe. (2009). 'Regional Clusters in a Global Industry: ICT Clusters in Canada.' *European Planning Studies* 17 (2) (February): 235–61.

Lundvall, Bengt-Åke. (2004). *Why the New Economy is a Learning Economy*. DRUID Working Paper No. 2004–01. Copenhagen.

Mallet, Jocelyn Ghent. (2002). *Silicon Valley North: The Formation of the Ottawa Innovation Cluster*. Information Technology Association of Canada. Available at: http://www.itac.ca (accessed 6 June 2005).

Martin, Ben. (2003). 'The Changing Social Contract for Science and the Evolution of Knowledge Production.' In *Science and Innovation: Rethinking the Rationales for Funding and Governance*, ed. Aldo Geuna, Ammon J. Salter, and W. Edward Steinmuller, 7–29. Cheltenham, U.K.: Edward Elgar.

Morgan, Kevin, and Claire Nauwelaers. (1999). 'A Regional Perspective on Innovation: From Theory to Strategy.' In *Regional Innovation Strategies: The Challenge for Less Favoured Regions*, ed. Kevin Morgan and Claire Nauwelaers, 1–18. St Crispins, U.K.: The Stationary Office.

Mowery, David C., Richard R. Nelson, Bhaven N. Sampat, and Arvids A. Zeidonis. (2004). *Ivory Tower and Industrial Innovation: University-Industry Technology Transfer Before and After the Bayh-Dole Act*. Stanford Business Books. New York: Cambridge University Press.

Nelles, Jen, Allison Bramwell, and David A. Wolfe. (2005). 'History, Culture and Path Dependency: Origins of the Waterloo ICT Cluster.' In *Global Networks and Local Linkages: The Paradox of Cluster Development in an Open Economy*, ed. David A. Wolfe and Matthew Lucas, 227–52. Montreal: McGill-Queen's University Press for the School of Policy Studies, Queen's University.

Nelson, Richard R., and Sidney G. Winter. (1982). *An Evolutionary Theory of Economic Change*. Cambridge, Ma.: The Belknap Press of Harvard University Press.

OECD, Science Technology Industry. (1999). *University Research in Transition*. Organization for Economic Co-operation and Development.

Pavitt, Keith. (1991). 'What Makes Basic Research Economically Useful?' *Research Policy* 20: 109–19.

Paytas, Jerry, Robert Gradeck, and Lena Andrews. (2004). *Universities and the Development of Industry Clusters*. Pittsburgh and Washington, D.C.: Center for Economic Development, Carnegie Mellon University and Economic Development Administration, U.S. Department of Commerce.

Pilieci, Vito. (2003). 'IBM Unveils Cutting-Edge Lab: Ottawa Facility a Boon to Graduate Students.' *The Ottawa Citizen*, 4 June.

Polanyi, Michael. (1967). *The Tacit Dimension*. New York: Doubleday Anchor Books.

Robinson, David. (2006). 'Changing Horses: Is the Innovation Strategy Working for the Mineral Sector?' In *Innovation, Science, Environment: Canadian Policies and Performance, 2006–2007*, ed. G. Bruce Doern, 149–68. Montreal: McGill-Queen's University Press.

Rosenberg, Nathan. (1990). 'Why Do Firms Do Basic Research (with Their Own Money)?' *Research Policy* 19: 165–74.

– (2003). 'America's Entrepreneurial Universities.' In *The Emergence of Entrepreneurship Policy: Governance, Start-Ups and Growth in the U.S. Knowledge Economy*, ed. David M. Hart, 113–37. Cambridge: Cambridge University Press.

Tornatzky, Louis G., Paul G. Waugaman, and Denis O. Gray. (2002). *Innovation U.: New University Roles in a Knowledge Economy*. Research Triangle Park, N.C.: Southern Policies Growth Board.

Walshok, Mary Lindenstein. (1995). *Knowledge without Boundaries: What America's Research Universities Can Do for the Economy, the Workplace, and the Community*. San Francisco: Jossey-Bass Publishers.

Warrian, Peter, and Celine Mulhern. (2009). 'From Metal Bashing to Materials Science and Services: Advanced Manufacturing and Mining Clusters in Transition.' *European Planning Studies* 17 (2) (February): 281–301.

Wolfe, David A. (2005). 'The Role of Universities in Regional Development and Cluster Formation.' In *Creating Knowledge, Strengthening Nations: The Changin Role of Higher Education*, Glen A. Jones, Patricia L. McCarney, and Michael L. Skolnik, 167–94. Toronto: University of Toronto Press.

Wolfe, David A., and Meric S. Gertler. (2004). 'Clusters from the Inside and Out: Local Dynamics and Global Linkages.' *Urban Studies* 41 (5/6) (May): 1071–93.

– (2006). 'Local Antecedents and Trigger Events: Policy Implications of Path Dependence for Cluster Formation.' In *Industrial Genesis: The Emergence of Technology Clusters*, ed. Pontus Braunerhjelm and Maryann Feldman, 243–63. Oxford: Oxford University Press.

Wolfe, David A., ed. (2003). *Clusters Old and New: The Transition to a Knowledge Economy in Canada's Regions*. Montreal: McGill-Queen's University Press for the School of Policy Studies, Queen's University.

Wolfe, David A., and Matthew Lucas, eds. (2004). *Clusters in a Cold Climate: Innovation Dynamics in a Diverse Economy*. Montreal: McGill-Queen's University Press for the School of Policy Studies, Queen's University.
– (2005). *Global Networks and Local Linkages: The Paradox of Cluster Development in an Open Economy*. Montreal: McGill-Queen's University Press for the School of Policy Studies, Queen's University.

12 Conclusions: Changing Symbiotic Research Relationships: Conflict and Compromise

G. BRUCE DOERN AND CHRISTOPHER STONEY

The purpose of this book has been to provide an integrated examination of the changing relationships between the federal government and Canada's universities as revealed through changes in federal research and innovation policies. The analysis has shown that the federal role vis-à-vis universities has increased in form and importance in the last decade through a new or changed set of research programs, processes, and institutions of engagement. In the basic literature on Canadian science and technology and related innovation policies, this kind of more integrated assessment has been absent.

This final chapter brings together our own overall views about why and how federal research and innovation policies have changed the relationships between the federal government and universities in Canada and the trends and dilemmas inherent in these reforms. Although our views draw on the analysis of our contributing authors, our aim is to highlight and connect the key arguments and themes that have emerged and review the need and opportunities for further changes as well as highlight the barriers, limits, and costs of these impacts and changing relationships.

The broad framework of the book set out in chapter 1 has facilitated a more detailed analysis and assessment of how the federal–university relationship has changed and the dilemmas this has brought into sharp focus between, for example, funding and performance, independence and accountability, academic values and commercialization, elitism and equity, centralization and decentralization. Key aspects of the framework's four analytical elements – high-level conceptual discourse, core values and ideas, policy instruments, and changes in institutions and governance have also been examined by our authors in their individual realms of federal S&T and innovation policies.

In this chapter the editors offer overall conclusions regarding: a) significant policy changes; b) complex 'symbiotic relationships,' initially between the federal government and universities, but also with business; the treatment of business relations and interests extends the conclusions somewhat beyond our federal–university focus per se but is necessary for our overall conclusions in part because federal S&T and innovation policies, as we have seen, have sought to establish closer partnered and levered relations between universities and business; c) the dynamic nature of these relationships amidst changing contexts, institutions, and policies; and d) the mutual adjustments needed regarding conflicts and compromises in values and interests.

Mirroring the book, this chapter also necessarily goes beyond the economic role of universities inherent in a focus on federal S&T and innovation policies. We maintain that policies aimed at realigning research should not be analysed and assessed purely in terms of their potential to drive the Canadian economy. Federal policies directed at reforming university research and outputs form part of the government's broader political and social agenda and must be seen in this broader context as well, especially to understand and explain the conflicts and compromises involved.

In many ways it is the political and social impacts of changing federal policy towards universities that have limited the amount of emphasis placed on the innovation and commercialization of university research. Universities have important community roles; academics are teachers as well as researchers; and the need for access, equality, and political support drains resources and impedes attempts to develop elite 'super universities' with mandates for 'world-class' research and innovation.

Federal S&T and Innovation Policies

The discussion of our analytical framework in chapter 1 showed the need to examine federal S&T and innovation policy in both linear and non-linear ways. Non-linear dynamics obviously mean that impacts also work in reverse through university researcher and institutional impacts on the federal government.

A summary of the main policy changes examined in detail in many of the earlier chapters reveals several important policy trends, and these will be used later to inform our discussion of symbiotic relationships, and of conflicts and compromise:

- Federal research funding was reduced in the 1990s due to federal deficit reduction strategies, but was also accompanied by some initial institutional experimentation through, for example, Networks of Centres of Excellence. Together these policies have helped influence later forms of partnered research.
- Funding from the late 1990s to the present was greatly expanded through a diverse and complex program architecture. The increased funding was obviously intended to stimulate research and innovation, while the newly developed architecture was part of the attempt to realign research priorities and reduce direct political control over funding.
- Each of the three granting bodies (NSERC, CIHR, and SSHRCC) has been pressured to reconfigure the proportion of its budget that goes to investigator-driven research as opposed to thematic and strategic research. This pressure has come both from federal departments and from entrepreneurial academic researchers. This is part of the attempt to develop strategic themes and coordinate individual efforts in specific areas. The effectiveness of this policy change is difficult to determine but it has necessarily produced winners and losers with regard to funding success.
- More complex boundary organization arrangements have emerged. The three granting bodies themselves are boundary organizations, as are universities in an overall sense. Boundary change processes have also occurred through peer review committees, which change over time as new realms of science, or equally, new interdisciplinary thematic areas of economic or social concern, wend their way in and among the interstices of established peer communities and journals and newly forming ones.
- Major new infrastructure funding has been provided via the CFI, but with contested impacts regarding how to view and assess research infrastructure and capacity. The CFI has had two overall regulatory impacts on universities: a) it has required universities to submit an overall strategic research plan, including many by universities which previously had no such plans; b) it has required levered money. Universities do not get infrastructure money unless they bring money from other public or private sources, and universities have widely varying capacities to raise such funding. Most of

the levered funding comes from the provinces rather than from business.
- Federal funding of research overhead costs has been belated, with the result that many universities pursuing research strategies as a priority dipped into teaching budgets. Though some overhead costs have now been covered this funding remains inadequate, as does overall understanding of the teaching impacts of research and innovation policies.
- Efforts and exhortations to enhance the commercialization of university research have increased, but have been accompanied by contested views of how best to accomplish this enhancement (indirectly through general human capital development and technology transfer and/or directly through patenting and spin-off companies), and the hoped for or expected revenue windfall and other related commercialization outputs have not been achieved. The real policy need may be to build a culture of research in the private sector, since the weakness in the commercialization process has been in the industry 'demand' side, not the university 'supply' side.
- Intellectual property in the form of patents has become increasingly important for universities and their technology transfer offices. However, the earnings from these patents/licensing arrangements have been small even when measured as gross revenues, let alone when net revenues are considered after deducting the costs of managing the IP.
- The policy-mandated increase in relocation of some government labs on university campuses favoured by the federal Conservative government needs greater understanding based on numerous arrangements already in place. There are good reasons to pursue the co-location model but it is by no means a panacea, and there are legitimate reasons as well to keep some key aspects of government science separate from university research.
- Increased partnered contract research has occurred between federal departments and university researchers in the natural, health, and social sciences. The horizontal dynamic of cooperation between institutions is simultaneously influenced by the hierarchical pressures of authority and accountability within each institution, thus creating both opportunities and barriers.
- More complex regulation of research ethics by federal granting councils and science-based agencies complements earlier professional

self regulation, but also bureaucratizes it considerably within universities and across research disciplines.
- Funding and networking impacts among the natural sciences, health research, and social sciences have been diverse, but the health sciences are doing better overall in garnering increased resources.
- There has been greater federal dependency on universities and university researchers for policy-related research and expertise. In addition to the increasing use of think tanks and party political advisors, this could be seen as part of the broader 'hollowing out' of the federal government's research and policy capacity.
- Involvement in regional/local cluster development has been extended in the name of innovation agendas, but with very limited federal control or influence. This shift highlights the tension between economic and social 'redistributive' criteria that pervades all Canadian policies and programs. While the reduction in direct political influence is regarded as a positive move, the development and productiveness of regional and local clusters is at best still patchy.

In focusing on these trends and impacts, several of our authors discuss the policies and values that govern institutions' actions and decisions, particularly their capacity to promote synergies and create conflicts. For, while universities, governments, and private industry have an overall common interest in engaging in productive and innovative research, the objectives of their research and the knowledge they wish to produce may not be and often are not the same.

The policies and values of the different institutions may produce different outcomes, namely academic understanding in the case of universities, practical public solutions in the case of governments, and commercial products and production processes in the case of private industry. They also result in the constantly shifting sands on which relationships, clusters, and partnerships are built.

As Doern analyses in chapter 4, the two core values of university research endeavour – the relative independence of researchers and peer review – must be reconciled with government values such as accountability for public funds and the need to demonstrate the relevance and usefulness of research outcomes. As we discussed in chapter 1, this tension can be expressed and understood as a classic example of the 'principal–agent' problem, which is by no means unique to government or to research. Similarly, these same university core values must somehow co-exist with private commercial goals and values as a result of private industry–university collaborations.

Given the public values at the heart of universities, one of Canada's few remaining state monopolies, Bird (chapter 7) finds it ironic that the university sector has been identified as the catalyst for Canadian private enterprise through the expansion of its commercially relevant research. In almost all other areas of government, neo-liberal reforms have been based on the premise that the public sector is inefficient and 'crowds out' private sector investment and expertise. While Canadian universities have been subjected to reform through the selective application of NPM principles, they have continued to attract significant increases in research funding, and are at the centre of strategies to improve Canada's record on innovation and technology transfer.

While the increasingly important research role of universities can be attributed partly to the hollowing out of the federal government's own research capacity over recent decades, it is the structural reliance of the state on economic growth that has contributed the most to the sustained and growing importance of universities. The generation of knowledge essentially means the generation of wealth and opportunity for the benefit of society as a whole through the linking of university research and development functions to governments' public policy and political goals and private companies' commercial goals. In many ways the transformation of our universities can be seen as part of a broader political ambition to create an enterprise state in Canada, one that is increasingly aligned with the interests, values, and outputs of the private sector. However, this does not imply that the interests of the state and the private sector are the same. Chapters 5, 6, and 11, for example, provide evidence that the interests of private industry are often different from the interests of the federal government in respect of universities.

At the same time as governments and private business have become increasingly dependent on universities, universities have become increasingly dependent on the resources that become available through their willingness to be influenced by government and private sector priorities. Access to adequate funding for research resources, both physical and intellectual, means that, as Kinder states in chapter 9, no institution can 'go it alone,' and this in essence is the basis for the symbiotic relationship. As we argue in this chapter, the irony of these growing symbiotic relationships is that, the closer universities are drawn into the commercial and public policy goals of their partners and collaborators, the greater the need for the core values of academic research to endure.

The book's chapters have explored many of the institutions that make up Canada's complex nexus of research capacity and resources – primarily the universities, federal government, and granting organizations, but also, tangentially, provincial governments, local communities, lobby groups, students, and the private sector – and the policies and inherent values which inform their actions and decisions.

In such an important, complex, and controversial policy arena, with so many vested interests, federal policies on research and innovation at universities are difficult to evaluate and appear to create as many unintended as intended consequences. Given the complexity and non-linear nature of the federal– university research nexus, the notion of cause-and-effect relationships in the context of stimulating innovation is also problematic for a number of our authors. Wolfe (chapter 11), for example, questions whether or not federal policy in the realm of cluster development really makes any predictable impacts on the nature and usefulness of research output. Certainly, according to Madgett and Stoney's assessment in chapter 6, the results of federal investment have, thus far, been disappointing in terms of raising levels of applied or commercially oriented innovation.

The framework developed for the book has established parameters for studying the complex relationships that have evolved between the institutions and stakeholders and the synergies and conflicts that result from the ever deepening interdependencies between them. By framing the analysis in this way, our analysis and assessment of Canada's research and innovation policies is able to adopt a nuanced and realistic lens through which to examine the evolving federal–university relationship.

In particular, this perspective helps us to understand why generating knowledge and innovation is considered to be a non-linear and long-term process that cannot be reduced to simple stimulus-response relationships. In terms of policy, several of our authors adopt this perspective to argue that increased federal investment in university research, while necessary, is not sufficient to achieve or maintain improved results.

On the other hand, the so-called 'winning conditions' that help stimulate innovation are not simply the result of serendipity or chance. The Canadian government may wish that increased funding in research would correlate more closely with tangible university and commercial outputs, but it also needs to support investment through consistent policies in a number of economic, social, and legal areas if it is to make a significant and sustained impact on research and innovation.

Too often federal policies lack coherence and consistency and appear to overlook the public values at the core of universities, such as the freedom of inquiry, and the demands placed upon the institutions as well as the academics who work in them. Faced with little management structure, few boundaries, and rapidly growing opportunities for research funding, private consultancy, and teaching buyouts, the dangers of commitment (and value) conflict, overload, and burnout are heightened significantly by the current policy trajectory.

Even if federal government policies were optimal in every respect, it would still not be sufficient to guarantee globally competitive innovation. As several of the chapters have discussed, the federal government is responsible only in part for university research and cannot, for example, control other levels of government. As Tupper outlines in chapter 2, both the provincial and federal levels of government play major roles in post-secondary education but, bewilderingly, without any formal coordination or formal delineation of these roles, it is an approach accurately described as 'uncoordinated entanglement.'

The federal government has even less influence over levels of private sector funding, or its capacity to work with and capitalize on the research produced. As we note in our introduction to the book, this begs an important question: to what extent is Canada's lack of innovation and technological transfer really a problem of demand as opposed to supply? Although this question takes us beyond the scope of the book it is clearly an important issue, and one that could have serious policy implications. In focusing on the university sector as the nexus between public investment and commercially viable innovations and inventions, there is a danger that the real problems in knowledge and technology transfer are not being addressed and that universities will be used as scapegoats for not bringing ideas to market by politicians, policymakers, and private corporations unwilling to criticize and overhaul their own roles in this complex chain.

Complex Symbiotic Relationships:
Beyond Federal–University Relations

To assess these varied impacts and related issues further and to draw some related conclusions, we need to discuss the notion of symbiotic relationships, namely the mutually advantageous association of two or more different institutions increasingly attached to each other or functioning partly within the other. We then look at the importance of achieving balance in the relationships between the federal government, universities, and commercial interests, particularly in respect of the values that underpin them, and consider as well the policy implications that impact on universities and innovation.

Through partnerships, centres of excellence, collaborations, and research clusters, governments and private industry seek to harness academic research capabilities and values in order to further their own research outcomes. The fundamental purpose of the relationship between universities and government, and indeed between universities and private industry, is for both parties to benefit from the mutual dependency (Rosenfeld 2006), but that is not always the outcome amid the interplays and tensions of the symbiotic relationship.

The issue of mutual benefit to the different members of the symbiotic relationship is a complex one, as it must take into account both the benefits and the costs to each institution. In particular, our interest is in the value of the symbiotic relationships for universities, and the ultimate price that might be paid as the dependencies on competitive government funding and private sector collaboration deepen. In effect, the close relationships between governments, universities, and private industry become a constantly evolving deal between the parties to achieve mutually satisfying outcomes without compromising their own distinct policies and values.

The benefits of this deal are multi-directional: if universities maintain their reputations and values as independent research institutions guided by academic freedom and integrity and critical peer review, the beneficiaries include not only the universities themselves and the academics and students working within them, but also the governments and private companies that seek to fund or partner their research endeavours. In other words, for governments and private industry, universities are not just another source of expertise; it is the inherent values of academic endeavour – intellectual freedom, integrity, scrutiny by peers – that make universities attractive and powerful allies in the pursuit of public goods, in the case of governments, and of commercial private goods, in the case of private industry. And, in turn, these values underpin the power and knowledge generated by the research partnership.

For research knowledge to be powerful, it must be credible, ethical, and unmitigated by conflicting demands on the producers of the knowledge. For governments, however, the need to win elections in order to remain in power necessarily forces both the political and the administrative wings of government into set time frames, electorally plausible justifications, and public accountability.

Similarly, for private industry, the need to make profits and defend investments to shareholders may influence decisions on how research should be conducted. Again, time frames and decisions are partially determined by the requirements of company accounts, speed to market, competitive pressures, and profitability. Collaboration with universities enables both governments and private industry to gain from the independence, intellectual integrity, and peer scrutiny that are associated with university research and academia in general to give their research outcomes credibility with the electorate (in the case of governments) and shareholders and customers (in the case of private industry).

However, the temptation for governments and private industry to align universities too closely to their own goals and pressures would threaten these qualities and thereby undermine academic research and analytical independence. In his commentary at the conference held to review the chapters in this book,[1] Dr Peter J. Nicholson stressed this central argument, as did James Turk of the Canadian Association of University Teachers. And while the government and private industry partners pay a price if university values are subjugated to their own, for the universities themselves and for individual academics, the price of compromising their academic values could include damage to personal reputations and to the standing of individual universities and of the Canadian university sector in general. Consequently, Canada's interest in establishing an internationally recognized academy of learning and research is both furthered by the symbiotic relationships among universities, governments, and private industry, and at the same time threatened by them. It cannot be in the interests of any of the institutions involved in the relationships to allow the costs to universities to be greater than the gains.

In particular, all the institutions involved need to consider how the symbiotic relationships can be leveraged for maximum effect without compromising the participants. If universities' research goals, values, and outcomes are, and are seen to be, too closely tied to governments' research objectives and values or private industry's commercial interests, the core

---

[1] Held at the National Arts Centre in Ottawa, 19–20 October 2006, the conference brought together authors, discussants, and participants from higher education and the federal government.

values of academic thought and endeavour must suffer. Independent, critical research, propelled not just by public policy goals or commercial viability, but also by academic integrity and enquiry, underpins the value of all collaborative outcomes of the symbiotic relationships.

Likewise, the interests of government researchers and private sector researchers are not the same as those of academic researchers – and to seek to unify these interests risks losing the inherent value of each. Generally, studies show that academics increasingly have less time to think, read, and consider wider research output, bringing into sharp relief the need for and risks to peer review and independent academic thinking. If academics fall into the model of consultant for government or the private sector, this risks compromising their academic integrity and independence and also simply takes up too much of their research capacity, in terms of time, energies, and resources. Finally, there is a danger that the agenda can be established so much by government funding bodies and industry that academics no longer set an independent research agenda.

The risks to wider society can be captured again in the threats to independence and freedom of academic thought and writing. In a free society, academics have a role in advising and criticizing policies of governments and quasi-public organizations. This freedom could be undermined if academics are too dependent on and too closely identified with government organizations or funding bodies; inevitably, university researchers will be reluctant to 'bite the hand that feeds them' by acting in turns as collaborator and as critic.

The authors have contributed studies which, like a jigsaw puzzle, assemble an overall picture of synergies and conflicts among institutions, policies, and values. The interplay among them, and the federal–university relationship in particular, underpins Canada's knowledge-based economy through innovation and the production of a knowledge-based labour force.

Dynamic Relationships: Changing Context, Institutions, and Policies

Additional complexity results from the fact that the economic, political, and social contexts within which the key institutions and players operate are not static but change continually. Consequently, relationships are in a constant state of flux as the synergies and conflicts are played out over the myriad projects and partnerships. The dynamic quality of the relationships is further complicated by changes in the institutions themselves, whose internal strategies, structures, priorities, and policies are themselves in a state of flux. This dynamic state reflects not only internal adjustments, changes, and barriers (including structural, financial, technological, and strategic changes within public, private, and university organizations), but also external changes in partner institutions.

There are two dynamic qualities that contribute to the complexities of the institutional framework. The first is that the institutional groupings that make up the apparently simple formula of the symbiotic relationships among governments, universities, and private sector organizations are comprised of different organizations with different priorities, powers, and capabilities. Within the groupings, there is both competition and cooperation in pursuit of their shared and individual goals.

The second is that, as the collaborative and financial relationships among universities, governments, and private industry become closer and more interdependent, these institutions become more directly affected by changes and developments in each other's policies and priorities. Thus, as Robert Best of the AUCC stressed in his conference comments,[2] changes in strategic direction or funding for one necessarily impinge on the operations of others in different jurisdictions. Clearly, government funding has a direct impact on partner organizations, but the reverse is also true – the capacity for universities and also private industry to produce quality research has an impact on the ability of government to generate solutions to policy problems. In order to examine the dynamic and codependent nature of these relationships, it useful to look at each core institution in turn.

Government

Changes in government priorities, the introduction of new public management techniques and ideas into government administration, and the emphasis on deficit reduction and realignment of financial resources from the federal government, followed by fiscal surpluses, had a dramatic effect on university resources and priorities and also on the shared jurisdiction of federal and provincial university policy and funding.

---

2 See n. 1, above.

Central to the relationships between Canadian universities and governments is a comprehensive understanding of the division of powers between the federal and provincial governments and the interests of these levels and the local level. Within Canada's federal model, constitutional responsibility for education lies with the provinces. Since 1976, when direct federal government grants to universities ceased, provincial governments have provided even higher portions of direct operating funding to universities as well as setting higher education policy in general. Because of the number of provincial jurisdictions involved in policy and direct grants, we have seen that there is no Canadian university 'system' per se. The provinces have jealously guarded their jurisdiction and resisted federal intervention in policymaking and direct funding.

The federal government supports universities through block transfer payments to the provinces, student assistance programs, and funding for university research. In recent years, the federal government has invested heavily in the last of these activities (chapter 1; Eastman, 2003). In particular, since 2004, the primary routes of the federal government's post-secondary spending have been via funding for students on the basis of both merit and need, and the promotion of academic scholarship through the diverse research grants, university chairs, and levered funding programs examined in earlier chapters.

The federal government's primary interest is the increase of research and development capability, but provincial governments also make their own contributions to R&D in universities. As Tupper's analysis in chapter 2 shows, the 1990s saw significant reductions in federal transfers to the provinces for post-secondary education. Cash transfers under the Canada Social Transfer Program are estimated to be 40 per cent lower than in 1992–93, adjusted for inflation and population growth. Although since 1998 funding levels have increased, underfunding remains a continuing problem for Canadian universities (Rosenfeld 2006). However, the issue of contention is for the most part not cost-sharing, but rather power-sharing. The federal government's contributions to the financing of provincial post-secondary education programs threatened the provinces' control and influence over the design and delivery of the programs. The federal government's interests are to meet federal targets or goals, and to establish national standards (for the purposes of national equity and regional development, and recently for global competitiveness). In addition, the federal government has sought to hold provincial governments more accountable for the expenditure of federal transfer funds and also to take some political credit for any achievements.

The federal model in Canada has resulted in universities retaining a more distant stance in terms of teaching and research, with both federal and provincial governments often failing to tie universities into explicit public goals and priorities, though at other times making such ties (chapter 1; Eastman 2003). The federal government lacks jurisdiction over education, and provincial governments have maintained an arm's-length control over universities, partly due to resistance from the universities and the AUCC, and partly due to the dependence of provinces on federal spending power through transfers and direct research funding and a lack of dynamic leadership on the part of the provinces, as argued by Morgan in chapter 3.

As previous chapters have shown, in the 1980s and 1990s, the many reforms introduced by Western governments into their oversight of universities sought to increase the autonomy of post-secondary educational institutions in order to 'marketize' their operations. By withdrawing government involvement from the inputs, university governance, and policies, the intention was to expose universities to the improving disciplines of the market and free them to enter into enterprising partnerships. However, the unintended effect of many of the measures was to impose greater controls over the universities through output measures and performance standards that in fact tied universities more closely into the state's public goals and policies. In New Zealand, Britain, and Australia, the reforms introduced under the auspices of the new public management model of public administration had far-reaching effects on universities' goals and priorities and became a key aspect of neo-liberal thinking in the area of education and academia (Eastman 2003), as well as in a myriad other areas of public policy (Borins 2002).

Ironically, Canada's federal structure prevented this from occurring to the same degree in Canadian universities, precisely because of the division of government powers and the resistance of provincial governments to federal interference. While it is true that new public management concepts never gained the same evangelical hold on thinking in public administration in Canada, the federal government and many provinces did undertake many new public management reforms. Under the Mulroney governments of the 1980s and the early1990s, and then the Chrétien Liberals in the mid-1990s, policies were characterized by the withdrawal of federal involvement and financial support as the debt repayment and budget controls became the order of the day until federal surpluses emerged in the late 1990s. Several provincial governments sought to introduce new public management measures more aggressively, most notably Ontario under the Harris government's 'Common Sense Revolution' in the mid to late 1990s. The federal model in Canada has proven to be flexible and adaptive, but it has also frustrated the explicit introduction of widespread public sector management reforms. It is the competing nature of this model that prompted Tupper, in chapter 2, to recall the quotation that federalism is both 'in flux and in cement.'

Universities

In a more general discussion of the changes in universities in Canada and their part in the symbiotic relationship with governments and private industry, it is tempting to consider all 91 university-type organizations in Canada as the same. However, the differences between them are considerable and their interests varied. Differences such as public and private universities, large and small ones, locations in big cities or small towns, with a local catchment or a wider catchment, specialized or general, are glaring, particularly in a country as big as Canada with such a dispersed population.

In this section, we consider the diversity of interests between universities in relation to stakeholders such as students, local communities and regions, and governments. In particular, an issue raised by several authors deserves highlighting, namely, that of the concentration of federal research and technology resources and funding on the country's most research-intensive universities. In his conference comments,[3] Professor David Cameron noted that such a concentration of resources has always been present, but by emphasizing this even more in recent years, the federal government exposes a major fault line between the promotion of research achievement through institutional differentiation, and the federal promotion of equality of opportunity across the whole country.

Canadian universities have an arm's-length corporate status, but they are funded by federal and provincial governments and are therefore deemed public. Canada's public universities' dependence on government funding distinguishes them from its private universities, which are few in number, mostly small, and religious-based (Eastman 2003).

---

3 See n. 1, above.

In addition, the complexity of the university institution is compounded by the different interests of the various faculties and departments, which necessarily have different capacities to forge partnerships and collaborations with government and private industry, depending on the relevance of their subject to contemporary research trends. Disciplines such as engineering and biomedical research, for example, lend themselves to collaborative efforts with government and private industry, and become increasingly dependent on the infrastructure, knowledge, and capabilities of private and government research.

In addition, some departments in the social sciences, such as public policy, political science, business administration, and international affairs, have what could be described as a hyper-symbiotic relationship with their 'partner' organizations (be they government departments, political institutions, or private sector organizations), that is, a relationship which is characterized not just by collaborations in joint research exercises or funding, but also an ongoing dependency on the 'partner' organization as the very subject of academic investigation. In such cases, academics who study and research political, government, or corporate organizations depend on access to and knowledge of these institutions' internal organization and actions.

Faculty in such disciplines need to maintain a healthy balance between the role of collaborator or consultant, and the role of academic critic. As previously discussed, the value of academic investigation in our society, as distinct from internal government or private sector investigation, lies in its autonomy from the subject – financially, institutionally, personally, and also morally. Universities perform a unique function within a democracy: to provide informed criticism of public or private policy that is independent and unbiased. Society depends on free speech and on the dissemination of balanced, informed, and unbiased debate on matters of political and economic import to generate productive, meaningful, democratic debate. Professors in relevant disciplines in the social sciences are much in demand by radio and television producers to speak on current affairs; their title and academic standing immediately signals to the audience that the opinion expressed can be trusted to be independent, well-informed, and unbiased.

This function could be compromised if academics are seen to be too closely identified with or dependent on a partner organization. Research funding awarded directly through government departments, for example, could make researchers reluctant to bite the hand that feeds them. Moreover, government departments could be expected to favour research proposals that do not inquire into sensitive policy areas or performance issues related to their own department.

Increasingly, partnerships between academics and private industry take the form of paid consulting contracts, which simultaneously further collaborative work between the institutions, increase academic access to the institution they are studying, and provide some external income for university professors. In response to this risk of compromised university values, Canadian universities are increasingly reviewing their policies on external relationships and professors' 'multiple roles' as university educators, academic researchers, consultants, and corporate directors. The phrase that is used to describe this risk is 'conflict of commitment' (Schmidt 2007). The Universities of British Columbia and Toronto have approved conflict rules to clarify the primary roles of teaching and research over external entrepreneurial activities. Other universities, such as the University of Ottawa, are following this lead. The conflict rules are designed to centralize and regulate the extra-curricular activities of professors through disclosure rules, standards of conflict of interest, and approval procedures (Schmidt 2007). The rules protect the reputation of the professors subject to them, and the standing and integrity of the institutions themselves.

Also, on an institutional level, research ethics boards exist within universities, private sector research organizations, and government organizations. The federal government's granting bodies have their own standing committees on ethics, such as the SSHRC's Standing Committee on Ethics and Integrity, and the CIHR's Standing Committee on Ethics (particularly important in the area of biomedical research with human subjects). In addition, universities have also established strong ethical committees, rules, and procedures in response to government and legal pressures. As Lavasseur points out in chapter 10, the trend from self-regulation to university and external regulation of ethics has created a complex labyrinth of ethics and bureaucracy that can impede research and innovation and increases the potential for competing and contradictory ethical frameworks.

According to Glen Jones, Associate Dean of Academic Affairs in the University of Toronto's Ontario Institute for Studies in Education (OISE), professors' research in collaboration with and funded by the private sector (or by government) can run the risk of compromising the teaching function of universities (cited in Schmidt 2007). In particular, he points to the new 'hybrid research projects' conducted through research funded by governments or private industry, which often involve the engagement of graduate students as paid researchers by the professor(s) running the project. Again, maintaining the delicate balance between professors' dual roles vis-à-vis students as, effectively, both employers and educators is essential to the fundamental teaching values of universities (Schmidt 2007) and needs to be framed as an ethical issue. This is not to say that such a balance should not be sought. Such hybrid, collaborative projects provide valuable opportunities for both professors and students alike to engage in relevant, up-to-date research and gain experience and knowledge. For students, it is a way of earning some income that may further their career aspirations and support and enhance their academic studies. Such valuable and worthwhile paid work is superior to students working sometimes long hours in low-paid, low-skill jobs to finance their studies.

In addition, it is worth returning to a central assumption of this book, namely, that universities play a critical role in the economic competitiveness of the country. Although there are valuable and sometimes lucrative spin-offs for universities as institutions (in the form of funding, infrastructure, and entrepreneurial opportunities), and for professors as individuals (in the form of research opportunities, sometimes lucrative consulting contracts, and funding), the paybacks for governments, private companies, and society at large are also considerable. Partnerships between universities and businesses generate 'cutting-edge' advances to the economy. Therefore, we can conclude that the advantages to be gained through collaborative work depend on the achievement and maintenance of the fine balances between shared and separate values.

A final point which needs to be considered is the issue of the concentration of government funding and research collaboration in a few research-intensive universities. Recently, federal research and technology transfer programs have funnelled resources into five or six 'super universities,' with the aim of building a concentration or cluster of research infrastructure and capability. Madgett and Stoney (chapter 6) consider Canada's ability to develop a research cluster on a geographic basis to be comparable to Silicon Valley near San Francisco or Route 128 near Boston. They argue that clusters do not happen by accident; funding for infrastructure, facilities, research chairs, scholarships, and partnerships and other collaborative efforts needs to be targeted to these few 'super-universities' through a tripartite relationship between government, the private sector, and the universities.

However, this entails the confrontation between competing values and ideas: such concentration of resources in a few universities means unequal funding across universities and necessarily relegates some universities, cities, towns, and regions to a second tier. As Scofield notes, a recent Conference Board of Canada study reinforces the view that, for Canada to compete in the global market, it needs to invest in education to produce a highly skilled workforce, and in research capacity to focus R&D resources. This would mean that Canada's long-held belief in the importance of equality of economic opportunity and development and social justice between peoples, regions, and provinces is potentially compromised by the value of economic competitiveness (Scofield 2007).

This belief forms a cornerstone of the Canadian political and economic system, at both the federal and provincial levels, through equalization payments, economic development programs, and social programs. For example, the federal government retains jurisdiction over the protection and promotion of the interests of Aboriginal peoples. At the same time, provinces have jurisdiction over education, health, and social welfare programs, all of which are predicated on equality of access to all Canadians. Therefore, the issue of institutional differentiation and the relegation of certain universities to a second tier raises an intractable question: in the national interest, which value takes priority? In the foreword to the Conference Board report, Janice Gross Stein argues that 'one size cannot fit all ... to educate young people, contribute to life-long learning, spark research and development, jump-start innovation and build global connections.' Government, private, and university resources need to be concentrated on those institutions and regions where the most gain can be achieved (Scofield 2007). The concentration of resources takes place on an institutional basis, as in the case of the University of Toronto, the University of British Columbia, and McGill, and also on a disciplinary level, as in the University of Saskatchewan and the University of Waterloo, where specialization takes place in subjects that are particularly relevant to or developed in that region.

The issue of concentration and specialization in research has far-reaching implications for equality of investment and opportunity across Canada's diverse regions and populations – for students, faculty, provinces, and regions. It is a prime example of where the interplay between institutions and values shows the complexity of policies aimed at generating knowledge and innovation. For example, the position being adopted by Scofield, the Conference Board, and others is predicated on the belief that, because highly skilled labour is now so globalized and mobile, the principle of subsidiarity – which in this context implies that investments should be made as close as is possible to where the benefits of the investment will be realized – no longer applies, and that real equality of economic opportunity now depends on specialized comparative advantage. If this is the case, then it brings into sharp focus the question of whether the concept of capacity building, discussed in chapter 5 in connection to the CFI, really accounts adequately for what the government thinks it is aiming to achieve.[4]

The Private Sector

While governments have a dual relationship with universities, one based on their responsibilities to citizens to produce top-quality postsecondary education and research capability, and the other based on their direct interests in universities as collaborators in research on a range of public policy problems, the interests of private industry in universities are simple: as collaborators in research which carries a market value that can be exploited.

The marketization of universities has been a gradual process over a period characterized by neo-liberalism and the adoption of 'new public management' (NPM) in the public and quasi-public sectors. NPM is predicated on the benefits of importing private sector management techniques and values into the public sector, consistent with a growing public mistrust of government and public sector decision-making. The apparent purity and simplicity of the profit motive in the private sector seemed to imbue private sector management with a clearer focus and a more customer-oriented approach, and this was seen as lacking in public sector management.

Reform of the public sector expanded to become the wholesale reinvention of the public sector into a 'post-bureaucratic' form, in which the core features and principles of the old bureaucratic framework – public interest, a focus on citizens and communities, collective interests, rules and standards of operation governing all decisions and expenditures, and a focus on procedure – are overtaken by a new set of principles embodied in NPM – such as service to 'clients,' quality and value, managerial rationality and empowerment, and a focus on government 'outputs' rather than on 'inputs' (Johnson 2002).

---

4 This point is based on the comments and insights provided by one of the book's anonymous reviewers.

As part of the NPM initiatives, universities were tied to more explicit performance standards and public goals of access, key skill development, and research output. However, the goal of the private enterprise is profit, as measured by the 'bottom line,' which establishes a clear set of customers. For the federal and provincial governments, and for the universities themselves, public goals are measured by more nebulous standards of provision of service and equality, where the 'customers' are often ill-defined groups of the public, including taxpayers at large. The private sector is oriented towards efficiency and competitiveness, whereas the public and quasi-public sectors must measure their success in terms of effectiveness of programs and services and equity amongst constituents.

Therefore, while collaboration with private industry is a valuable source of resources, expertise, and capacity, the clear differences in motivation between private and public/quasi-public organizations, and private and public goods, reinforce the need for a fine balance between collaboration and autonomy on the part of the universities. The transformations of the past decade for business as a result of technology and the knowledge economy are held to be greater than the effects of the industrial revolution; for business, the need for universities to deliver skilled labour and research capability are paramount (Oblinger and Verville 1998).

Moreover, either through matched funding or collaborative partnerships with universities on an institutional level, or through consultant contracts with professors on an individual level, private industry reaps the benefits of academic knowledge and the credibility and standing of academic research. However, this again reinforces the need for a fine balance so that private industry may benefit from the collaboration without compromising the academic partner's values, and vice versa.

Mutual Adjustment: Conflicting Values and Interests

While policies can be defined as the formal, tangible, written rules and directions of an institution, its values are the foundation stones of the attitudes, beliefs, and assumptions that often remain unwritten and informal, but play a central role in the functioning of the institution. The clash between public, quasi-public, and private values is examined throughout the book and specifically in discussions concerning the impact of new public management.

In keeping with NPM ideas, the model of collaborative research among government, universities, and private industry is built on attempts to establish the post-bureaucratic model, which is built on a fundamental separation of the functions of policy formulation and of implementation. Thus, governments increasingly seek to empower universities and private organizations to implement government policies and priorities on the assumption of shared values and complementary functions. However, in reality such assumptions cannot be fully realized and values compete for priority within the partnership.

Governments, as we have seen, have obligations derived from the democratic system of government which require them to account for the expenditure of public monies and to serve the national interest through uniformity of service delivery and equality in national coverage. These values, as with all forms of alternative service delivery used by federal and provincial governments, conflict with the empowering of other institutions to implement government policy.

Similarly, the Conference Board report referred to above argues that Canada's culture of equality across all provinces seriously undermines the country's ability to compete in the global economy. The report advocates the concentration of investment resources in certain key regions and institutions in order to 'create excellence and spawn more wealth,' rather than pursue public policies of equality (Scofield 2007).

According to LaPointe (chapter 8), these sometimes fundamental differences produce a 'language of barriers' that both reflects and produces difficulties in reconciling institutional values in the pursuit of shared aims and that ultimately undermines trust. Consequently, accommodating the different values of institutions is key because they create the cultural and psychological conditions for collaborative relationships and underpin the concept of trust.

In identifying three pillars of collaboration, LaPointe identifies trust, along with time and outcomes, as a necessary and key ingredient for the sharing of knowledge between institutions and individual researchers. On an individual level, the psychological aspect of true partnership is built on the foundations of trust and mutual reliance, which in turn are supported by the organizational culture of their institutions.

From this perspective, it becomes clear that the values that come into play in any collaboration between different institutions must have a considerable influence on trust and the success of the outcomes. If these values clash, then trust and collaborative efforts will be compromised, perhaps to the extent that longer-term relationships become fruitless and counterproductive. Given the different interests and values that underpin the motives of the institutional players involved, it is not difficult to see why such scenarios occur.

Trust is an emotive concept that is crucial to interpersonal collaborative relationships and requires a degree of mutual adjustment in respect of each participant's values and priorities. This is necessary to create a framework which facilitates knowledge and information-sharing and reciprocity concerning resources, ideas, and results. Through such frameworks, trust becomes a cornerstone of knowledge production and transfer, as power is acquired and shared to produce synergy, innovation, and value.

Critically, trust cannot be forced, however, and can only be built up over time through shared experiences of collaboration. Nevertheless, attempts to develop such relationships are constantly impeded by conflicting and contradictory forces inherent within the system. First, for example, competition between individuals, universities, and different institutions within the partnerships is heightened by the rivalry for funding, infrastructure, and expertise. The irony of the notion of the knowledge economy is that a market economy is built on personal and institutional competition that encourages the exploitation of knowledge, resources, and capital for one's own benefit and withholding it from others.

There is unquestionably a dynamic, productive quality to competition, which powers knowledge creation and innovation. However, competition also can be dysfunctional and counterproductive if funding applications and commercial goals make knowledge-sharing unattractive. Between private companies there can be competition to develop superior products and be first-to-market. Similarly, there is competition between universities (as discussed, between the big and small universities) for funding and academic personnel.

Second, for academics themselves, there is growing evidence of increased pressure on their time, resulting in a 'fog of fatigue and culture of overwork' (Menzies and Newson 2006). Recent research by the *Journal of Higher Education: Academic Matters* on academic workload revealed that a majority of academics are not able to read 'as deeply and reflectively' as they want, or 'as broadly and interdisciplinarily' (Menzies and Newson 2006). This suggests that there is a clash between the amount of work academics are engaging in – including teaching, administrative, media, community, and research work – and the depth and breadth in which they are able to do it. This also relates specifically to their capacity for networking and partnership-building, which requires time and commitment to establish, manage, and develop. As stated earlier, this means that excessive demands on academics' time undermine the value of academic endeavour for all parties.

For governments in particular, universities and academics are now an important resource in the implementation of public goals. Their autonomy, credibility, and integrity as researchers and critics make them valuable partners for governments. However, complex relationships between the key players are inevitable when public goals (in the form of solutions to public problems), commercial goals (in the form of private products), and university goals (in the form of funding, academic standing, and attracting students) come together in a single pursuit. Recognition and management of these conflicting values and demands will ultimately have a great bearing on the effectiveness of federal policy and the outcomes produced.

Conclusion

In the context of the changing federal–university relationship regarding research and innovation policies, a number of key issues ultimately emerge. The first is the continuing generation of research knowledge and the need for current knowledge and skills to update and maintain the knowledge advantage. The speed and competition in research and knowledge generation mean that any competitive advantage and commercial value are quickly expended and the demands and pressures on researchers and institutions increase.

Universities have a vital role to play in research and innovation, but we need to remember that they are constrained by resources, by the demands on their academics, by the dual role of knowledge production and knowledge transfer (in particular teaching), values, and, in particular, by the need to remain independent.

Furthermore, the pursuit of research and knowledge engenders a tension between trust and the sharing of that research and knowledge and the competition between universities, companies, different institutions, and individual researchers and teams of researchers. Power means that research and knowledge becomes a commodity that can be withheld, corrupted, sold, and traded. For viable and productive research to take place within collaborative institutional arrangements, research and knowledge needs to be shared within long-lasting, trusting, complementary partnerships.

The various balancing acts to which all the institutions and individuals contribute must ensure that this tension remains dynamic and stimulating rather than destructive and counterproductive. Clearly, the ideal is a balance of collaboration and mutual dependency with independence and adherence to institutional and personal values. However good such a balance may be, though, conflicts of institutional goals, values, and priorities are inevitable and accommodation is required on all sides. A key measure of the success of a research collaboration or partnership is the outcomes it produces; yet the outcomes desired by universities, governments, and industry are necessarily very different.

Good relationships between these institutions are not simply desirable, but essential to stimulate and promote profound, relevant research. However, the mutual dependency and shared outcomes cannot be seen as resolving the inherent conflicts of institutional and personal goals and values. Not only are the symbiotic relationships not able to resolve the conflicts, but indeed they should not – for if such conflicts of goals and values are resolved, it can only be if one institution's are subjugated in favour of another's.

The network of institutions and values that comprises the contemporary research and development (R&D) community opens up new avenues of collaboration and cross-fertilization that involve in various ways a myriad of interested groups and organizations, and increase the multi-directional flow of knowledge that fuels the knowledge economy. But this cannot and should not be at the expense of independent academic enquiry. To do so would be to 'kill the goose that lays the golden egg,' in the sense that it would denude the universities of their most valuable and desirable qualities, both for the internal stakeholders in the university community (such as students, faculty, university governors) and for external partners, funders, and collaborators.

As the federal government attempts to improve Canada's record of research and innovation it is limited by the fact that it is only one of many players, and that stimulating research and innovation is a nonlinear process; by its need to balance economic, political, and social criteria; by the demands of regional and national priorities; and, of course, by the constraints imposed by the national and global economies.

In addition, there are concerns that federal policies, driven by the imperative of economic growth, tend to favour natural and applied sciences at the expense of social sciences, many of which have an important contribution to make to society as well as to the transfer of knowledge to the market. Also, it is not clear that policies aimed at promoting collaborative interdisciplinary research are producing the expected increases in innovation, and may in any case be undermined by increased competition for funding, complex administration, and the unsustainable demands placed on academics.

Notwithstanding these concerns and constraints, this book has examined ways in which the federal government can play an important role in stimulating knowledge and innovation through appropriate policies and initiatives. While funding, co-location of laboratories, creation of arm's-length bodies, and so on can be seen to have the most direct impact on university activities, the government's broader role in society is very likely more important and should not be overlooked by policymakers.

As Canada's national government, the federal government is uniquely placed to coordinate the regions and cooperate with the provinces to establish priorities and negotiate reform. Through regulation, taxation, and investment it is able to create incentives, influence demand, and pursue the 'winning' economic, political, social, and legal conditions that underpin research, knowledge transfer, and innovation.

Given that Canada is slipping internationally in terms of its research and development, faces declining productivity, and is increasingly dependent on exports of its natural resources, few would argue against the federal government taking steps to invest in universities and knowledge production. However, research and reflection presented in this book confirm that the process of stimulating innovation is complex and does not respond well to top-down command and control directed by federal government.

Knowledge transfers from universities to industry and government are part of an iterative process, personalized through trust relationships and localized in regional and institutional clusters. Policies and investments that strengthen rather than undermine these aspects of the symbiotic relationship will provide a framework for facilitating and aligning competing values and priorities. Ultimately, it is by encouraging and supporting academics in the pursuit of personal and institutional goals that the federal government can provide the best opportunity for stimulating knowledge, innovation, and economic growth.

REFERENCES

Borins, Sandford. (2002). 'Transformation of the Public Sector: Canada in Comparative Perspective.' In *Handbook of Public Administration*, ed. Christopher Dunn, 4–5. Don Mills, Ont.: Oxford University Press.
Eastman, Julia. (2003). 'Strategic Management of Universities?' In *Canadian Society for the Studies of Higher Education: Professional File*, no. 24 (Fall).
Johnson, David. (2002). *Thinking Government: Public Sector Management in Canada*. Peterborough, Ont.: Broadview Press.

Menzies, Heather, and Janice Newson. (2006). 'No Time to Think?' *Journal of Higher Education: Academic Matters* (Winter): 12–15.
Oblinger, Diane, and Anne-Lee Verville. (1998). *What Business Wants from Higher Education*. American Council on Education Oryx Press Series on Higher Education.
Rosenfeld, Mark. (2006). 'Academic Freedom and Public Policy.' *Journal of Higher Education: Academic Matters* (Fall): 32.
Schmidt, Sarah. (2007). 'Moonlighting Puts Professors in Conflict of Commitment.' *Ottawa Citizen*, 5 March 2007, A2.
Scofield, Heather. (2007). 'Equality "Myth" Seen Holding Canada Back.' *Globe and Mail*, 18 January 2007, B5.

Appendix: Key Research and Innovation Data

GERD   Gross domestic expenditures on research and development
GOVERD   Government intramural expenditure on research and development
BERD   Business enterprise expenditure on research and development
HERD   Higher education expenditure on research and development

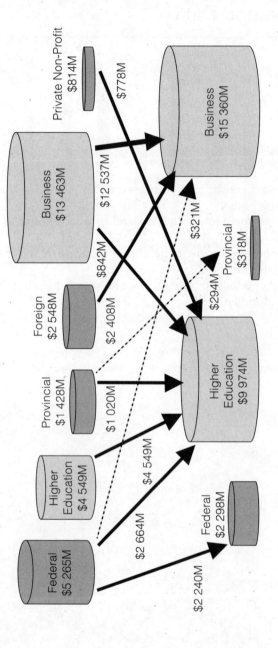

Figure 1: Major Flows of R&D Funding in Canada, 2006*

*Only flows higher than $150M are shown in the chart.
Source: Statistics Canada, *Science Statistics*, Vol. 31, No. 8, December 2007 and CANSIM Table 358 0001.

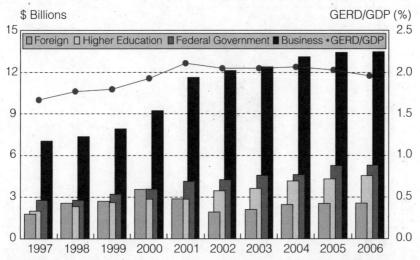

Figure 2: Canada's GERD by Major Source of Funds, 1997 to 2006

Source: Statistics Canada, *Science Statistics*, Vol. 31, No. 8, December 2007.

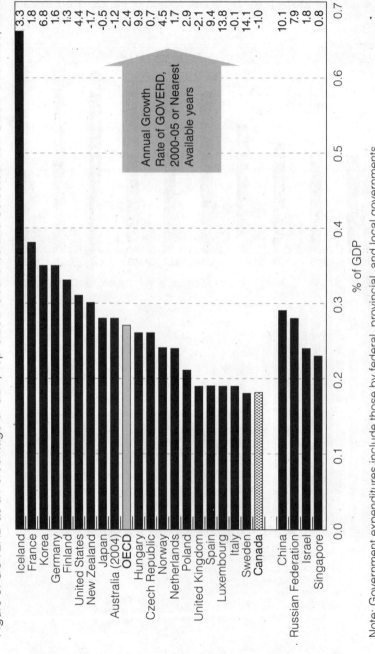

Figure 3: GOVERD as a Percentage of GDP, Top OECD Countries and Selected Non-OECD Countries, 2005

Note: Government expenditures include those by federal, provincial, and local governments.
Source: OECD, *Main Science and Technology Indicators 2007/2*, November 2007.

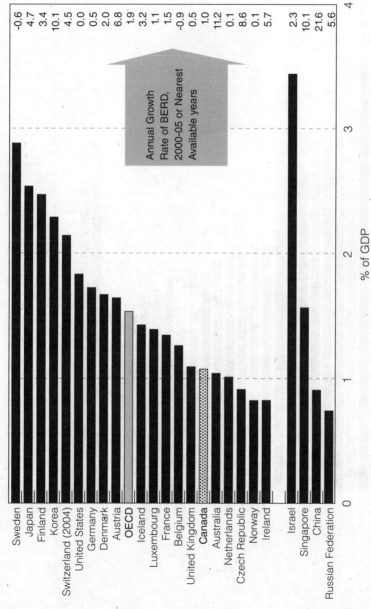

Figure 4: BERD as a Percentage of GDP, Top OECD Countries and Selected Non-OECD Countries, 2005

Source: OECD, *Main Science and Technology Indicators 2007/2*, November 2007.

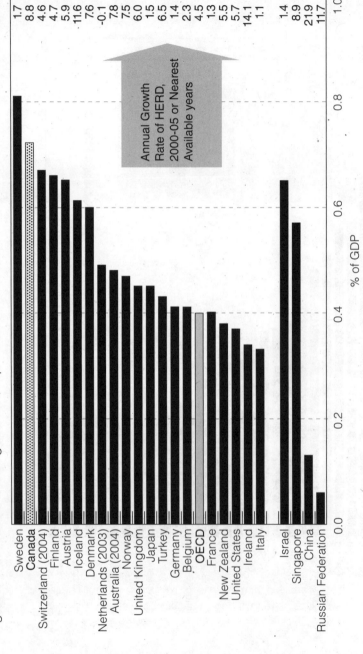

Figure 5: HERD as a Percentage of GDP, Top OECD Countries and Selected Non-OECD Countries, 2005

Source: OECD, *Main Science and Technology Indicators 2007/2*, November 2007.

Figure 6: Human Resources in S&T as a Percentage of Total Employment, Selected OECD Countries, 2006

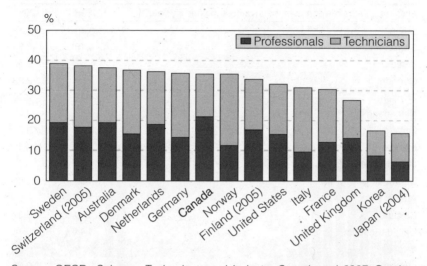

Source: OECD, *Science, Technology and Industry Scoreboard 2007*, October 2007.

Figure 7: Selected Commercialization Outputs of University Research by Scale of Income Generated,* 2005

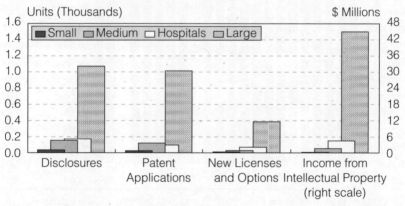

* Income is from sponsored research. Small is less than $25 million; medium is $25 million to $79 million, and large is $80 million or more. Hospitals associated with universities were included to ensure consistency with the aggregate figures. Source: Statistics Canada, *Innovation Analysis Bulletin*, Vol. 9, No. 1, Cat. No. 88-003-XIE.

Table 1
Triadic Patent Families:[1] As Percentage of GDP[2]

According to the residence of the inventors, by priority year:	1996	2005
Japan	3.43	4.15
Switzerland	4.18	2.85
Korea	0.57	2.63
Germany	2.89	2.33
Sweden	4.12	2.10
Netherlands	2.47	1.79
Finland	3.71	1.55
France	1.68	1.18
United States	1.77	1.17
Denmark	1.88	1.06
Austria	1.12	0.98
Belgium	1.59	0.87
United Kingdom	1.44	0.70
Canada	0.68	0.66
Luxembourg	0.94	0.65
New Zealand	0.52	0.53
Australia	0.58	0.50
Norway	0.71	0.44
Iceland	1.25	0.38
Italy	0.59	0.38
Ireland	0.44	0.31
Hungary	0.36	0.16
Spain	0.15	0.14
Czech Republic	0.10	0.06
Portugal	0.03	0.04
Greece	0.08	0.03
Slovak Republic	0.03	0.03
Poland	0.04	0.02
Turkey	0.04	0.02
Mexico	0.03	0.01

1. Patents all applied for at the EPO, USPTO, and JPO. 2005 figures are estimates.
2. Gross Domestic Product (GDP), billion 2000 USD using purchasing power parities.
Source: OECD, Main S&T Indicators, 2007/2.

Contributors

Malcolm G. Bird obtained his doctorate in 2008 in the School of Public Policy and Administration at Carleton University. His PhD thesis was on public ownership and liquor control policy.

G. Bruce Doern is Chancellor's Professor in the School of Public Policy and Administration at Carleton University and holds a joint Research Chair in Public Policy in the Politics Department at the University of Exeter. He also served as the Director of the Carleton Research Unit on Innovation, Science and Environment (CRUISE) from 2002 to 2007.

Jeffrey S. Kinder is a PhD candidate in the School of Public Policy and Administration at Carleton University. His research interests include science, technology, and innovation policies and institutions. He is the co-author with Bruce Doern of *Strategic Science in the Public Interest: Federal Laboratories and Science-Based Agencies* (University of Toronto Press, 2007).

Russell LaPointe is a PhD candidate in the School of Public Policy and Administration at Carleton University. His research interests include Aboriginal policy and treaty-making, and S&T and innovation policies and institutions.

Karine Levasseur obtained her doctorate in 2008 in the School of Public Policy and Administration at Carleton University. Her research interests include state–civil society relationships, family policy, and government regulation.

Débora Lopreite obtained her doctorate in 2009 in the School of Public Policy and Administration at Carleton University. Her research interests include democratic governance, comparative public administration, and innovation policies and institutions.

Paul J. Madgett is a PhD candidate at the University of Georgia. His research interests centre on higher education policy and institutions.

Clara Morgan obtained her doctorate in 2008 in the School of Public Policy and Administration at Carleton University. Her doctoral research was on the OECD's education policies. She is interested in all aspects of education policy and governance at the elementary, secondary, and higher systems of education, as well as the internationalization of education.

Joan Murphy obtained her doctorate in 2008 in the School of Public Policy and Administration at Carleton University. Her research interests include health policy and research and innovation policy. Her PhD thesis was on the formation and evolution of the Canadian Institutes of Health Research.

Christopher Stoney is Associate Professor in the School of Public Policy and Administration at Carleton University, where he teaches MA courses on federal and local government, public sector management, organizational theory, strategy, and urban sustainability.

Allan Tupper is Professor of Political Science at the University of British Columbia. His major current research interests are in post-secondary education, government ethics, and administrative reform. He is the co-author with Tom Pocklington of *No Place to Learn: Why Universities Aren't Working* (University of British Columbia Press, 2002).

David A. Wolfe is Professor of Political Science at the University of Toronto and Co-Director of the Program on Globalization and Regional Innovation Systems at the Munk Centre for International Studies.